SEASONS OF LIFE
STUDY GUIDE

by

Richard O. Straub
University of Michigan-Dearborn

Worth Publishers, Inc.

From
The Annenberg/CPB Collection

Seasons of Life Study Guide
by Richard O. Straub

Copyright © 1990 by The Regents of The University of Michigan.
All rights reserved.

This *Study Guide* is part of a full college course, which also includes
five one-hour television programs; 26 half-hour audio programs;
The Developing Person Through the Life Span, second edition, by
Kathleen Stassen Berger (Worth Publishers, Inc.); and a *Faculty
Manual and Test Bank.* For information about licensing the course,
purchasing video and audiocassettes or course print components,
call 1-800-LEARNER, or write the Annenberg/CPB Project, P.O.
Box 1922, Santa Barbara, CA 93116-1922.

Other materials based on **Seasons of Life** include 20 six-minute
audio modules and a companion book, *Seasons of Life,* by John
Kotre and Elizabeth Hall (Little, Brown and Company). The book is
available by calling 1-800-LEARNER and in most bookstores.

Funding for **Seasons of Life** is provided by the Annenberg/CPB
Project.

ISBN: 0-87901-416-4

Printing: 4   3   Year: 94   93   92   91

Worth Publishers, Inc.
33 Irving Place
New York, NY 10003

PREFACE

Seasons of Life is an introductory telecourse in life-span development—an exploration of the fascinating biological, social, and psychological changes that occur from the beginning of life to its end. In *Seasons of Life* you'll meet dozens of people in all stages of life and hear the views of nearly 50 leading social scientists.

Each of the twenty-six lessons of the telecourse consists of a thirty-minute audio program hosted by psychologist John Kotre; a chapter from *The Developing Person Through the Life Span, second edition,* by Kathleen Stassen Berger; and an assignment in this *Study Guide.*

Seasons of Life also features five one-hour television programs hosted by David Hartman, formerly of ABC's "Good Morning, America." Each program looks at a particular stage, or "season," of life. Program One introduces the "clocks" that influence biological, social, and psychological development throughout life and explores development during the first six years. Program Two covers development during childhood and adolescence (ages 6–20), and Program Three looks at early adulthood (ages 20–40). Program Four focuses on middle adulthood (ages 40–60) and Program Five on late adulthood (ages 60+).

This *Study Guide* is designed to help you accomplish the goals of this course and to evaluate your understanding of the telecourse material. Your instructor will inform you of the lessons and assignments to be completed. At the end of the course you will be asked to complete a Television Term Project. The purpose of the project is to help you to integrate the material presented in the audio, video, and textbook components of the telecourse and to interpret your own experiences in view of what you have learned.

This *Study Guide* also includes a section, "How to Manage Your Time Efficiently and Study More Effectively" (page viii), which provides information on how to use the *Study Guide* for maximum benefit. It also offers additional study suggestions for time management, effective note taking, evaluation of exam performance, and ways to improve your reading comprehension.

We are grateful to everyone who has contributed to this project. Special thanks go to our core group of advisors: Urie Bronfenbrenner of Cornell University, David Gutmann of Northwestern University, Bernice Neugarten of the University of Chicago, Anne Petersen of Penn State University, Alice Rossi of the University of Massachusetts–Amherst, and Sheldon White of Harvard University. Jack Mitchell of WHA at the University of Wisconsin provided much wise counsel regarding audio programming. Janet Whitaker of Rio Salado Community College was especially helpful in the design of our supplementary materials. Margie Moeller of WQED wrote excellent descriptions of the people in the television programs. Wilbert McKeachie and Elizabeth Douvan of the University of Michigan offered help close to home, and the staff of Worth Publishers supplied patience and professionalism. Major funding for *Seasons of Life* was provided by the Annenberg/CPB Project.

In preparing the materials that make up the *Seasons of Life* project, we were guided by a simple idea: that by examining the stories people tell about their lives, we can come to an understanding not only of these lives but of the life cycle itself, and of its "seasons." We hope your journey through the telecourse experience will be an enjoyable one and provide you with information that enhances your insight into your own life story and those of others.

John Kotre, Ph.D.
Professor of Psychology
    University of Michigan–Dearborn
Project Director, *Seasons of Life*

Richard O. Straub, Ph.D.
Associate Professor of Psychology
    University of Michigan–Dearborn

TABLE OF CONTENTS

The Seasons of Life Study Guide

This study guide has many features that will help you study more effectively. Part I of this introduction describes these features. Part II provides suggestions that can help you to use your study time more effectively.

PART I   FEATURES OF THIS STUDY GUIDE

THE LESSONS

Each Study Guide lesson consists of nine sections designed to help direct your study activities. After a while, you will discover which sections are the most helpful for you, and you can concentrate on them.

Orientation   This section provides an overview of the lesson, highlighting important themes, placing facts in context, and integrating the audio program with the textbook. It also introduces the experts and stories that appear in each audio program and helps you relate its content to important ideas covered in other lessons.

Lesson Goals   These goals—typically four or five for each lesson—identify the major themes of each lesson. They are drawn from both the textbook and the audio program.

Audio Assignment   This section contains several specific questions drawn from the important facts and concepts covered in the lesson's audio program. Once you have finished listening to the program, try to answer the questions in your own words. Completing this section will help you to identify both concepts you have grasped and ones you may need to review by replaying portions of the program.

Textbook Assignment   This section contains several specific questions drawn from important facts and concepts presented in the textbook chapter. Once you have finished reading the chapter, try to answer the questions in your own words. Completing the questions will help you to identify those points that you understand, as well as those you may need to review.

Testing Yourself   This section consists of 15 multiple-choice questions drawn from both the audio program and textbook chapter. They should be answered only after you have listened to the program, read the chapter, and completed the audio and textbook assignments. Correct answers, along with explanations, are provided at the end of the chapter. If you miss a question, read the explanation and, if you need to, review the appropriate text pages or portion of the audio program.

Exercise   This section contains a short assignment that you are to complete and return to your instructor. The exercise will help you make meaningful connections between the lesson content and your own life story. In some cases the exercise involves gathering information from a relative or friend. In others, it is based on applying important lesson concepts to your own experiences.

Lesson Guidelines   Organized into separate sections for the audio program and the text, this section condenses and restates the main points covered in the lesson. It will help you to evaluate your answers to questions in the "Audio Assignment" and "Textbook Assignment" sections of the Study Guide.

Answers to Testing Yourself   This section can be used to evaluate your performance on the questions in the "Testing Yourself" section of the Study Guide. For each question, the correct answer is given, along with the question's source—whether material in the audio program or the text. For questions based on the text, page references are given where the answer can be found.

References   In this section, you are referred to journals, magazine articles, and books that contain additional information on topics discussed in the lesson.

THE TELEVISION TERM PROJECT

The Television Term Project consists of 25 essay questions designed to integrate the audio, video and print components of the telecourse. The questions are divided into two categories. For each television program, there are four questions which should be answered soon after the program is viewed. For the entire series, there are five additional questions that cover major themes and help you to make meaningful connections between the telecourse and your own life experiences.

KEEPING TRACK

The following grid will help you keep track of your progress through the *Seasons of Life* telecourse. Your actual assignments, of course, will depend on your instructor.

Lesson	Exercise Score	Quiz Score
1		
2		
3		
4		
5		
6		
7		
8		
9		
10		
11		
12		
13		
14		
15		
16		
17		
18		
19		
20		
21		
22		
23		
24		
25		
26		
Total		
Average		

Exercise average	
Quiz average	
Mid-term exam	
Final exam	
Television term project	
Overall Average	
Final Grade	

Part II HOW TO MANAGE YOUR TIME EFFICIENTLY AND STUDY MORE EFFECTIVELY

Students who are new to college life or who are returning after a long absence may be unsure of their study skills. Suggestions for making the best use of your time and improving your skills may be found in the following section of the Study Guide. These suggestions will help you not only with *Seasons of Life*, but also with many other of your college courses.

How effectively do you study? Good study habits make the job of being a college student much easier. Many students who could succeed in college fail or drop out because they have never learned to manage their time efficiently. Even the best students can usually benefit from an in-depth evaluation of their current study habits.

There are many ways to achieve academic success, of course, but your approach may not be the most effective or efficient. Are you sacrificing your social life, or your physical or mental health, in order to get A's on your exams? Good study habits result in better grades and more time for other activities.

EVALUATE YOUR CURRENT STUDY HABITS

To improve your study habits, you must first have an accurate picture of how you currently spend your time. Begin by putting together a profile of your present living and studying habits. Answer the following questions by writing "yes" or "no" on each line.

_____ 1. Do you usually set up a schedule to budget your time for studying, recreation, and other activities?

_____ 2. Do you often put off studying until time pressures force you to cram?

_____ 3. Do other students seem to study less than you do, but get better grades?

_____ 4. Do you usually spend hours at a time studying one subject, rather than dividing that time between several subjects?

_____ 5. Do you often have trouble remembering what you have just read in a textbook?

_____ 6. Before reading a chapter in a textbook, do you skim through it and read the section headings?

_____ 7. Do you try to predict exam questions from your lecture notes and reading?

_____ 8. Do you usually attempt to paraphrase or summarize what you have just finished reading?

_____ 9. Do you find it difficult to concentrate very long when you study?

_____ 10. Do you often feel that you studied the wrong material for an exam?

Thousands of college students have participated in similar surveys. Students who are fully realizing their academic potential usually respond as follows: (1) yes, (2) no, (3) no, (4) no, (5) no, (6) yes, (7) yes, (8) yes, (9) no, (10) no.

Compare your responses with those of successful students. The greater the discrepancy, the more you could benefit from a program to improve your study habits. The questions are designed to identify areas of weakness. Once you have identified your weaknesses, you will be able to set specific goals for improvement and implement a program for reaching them.

MANAGE YOUR TIME

Do you often feel frustrated because there isn't enough time to do all the things you must and want to do? Take heart. Even the most productive and successful people feel this way at times. But they establish priorities for their activities and they learn to budget time for each of them. There's much in the saying, "If you want something done, ask a busy person to do it." A busy person knows how to get things done.

If you don't now have a system for budgeting your time, develop one. Not only will your academic accomplishments increase, but you will actually find more time in your schedule for other activities. And you won't have to feel guilty about "taking time off," because all your obligations will be covered.

Establish a Baseline

As a first step in preparing to budget your time, keep a diary for a few days to establish a summary, or baseline, of the time you spend in studying, socializing, working, and so on. If you are like many students, much of your "study" time is nonproductive; you may sit at your desk and leaf through a book, but the time is actually wasted. Or you may procrastinate. You are always getting ready to study, but you rarely do.

Besides revealing where you waste time, your diary will give you a realistic picture of how much time you need to allot for meals, commuting, and other fixed activities. In addition, careful records should indicate the times of the day when you are consistently most productive. A sample time-management diary is shown in Table 1.

Plan the Term

Having established and evaluated your baseline, you are ready to devise a more efficient schedule. Buy a calendar that covers the entire school term and has ample space for each day. Using the course outlines provided by your instructors, enter the dates of all exams, term paper deadlines, and other

TABLE 1   Sample Time-Management Diary

Behavior	Monday Time Completed	Duration Hours: Minutes
Sleep	7:00	7:30
Dressing	7:25	:25
Breakfast	7:45	:20
Commute	8:20	:35
Coffee	9:00	:40
French	10:00	1:00
Socialize	10:15	:15
Videogame	10:35	:20
Coffee	11:00	:25
Psychology	12:00	1:00
Lunch	12:25	:25
Study Lab	1:00	:35
Psych. Lab	4:00	3:00
Work	5:30	1:30
Commute	6:10	:40
Dinner	6:45	:35
TV	7:30	:45
Study Psychology	10:00	2:30
Socialize	11:30	1:30
Sleep		

Prepare a similar chart for each day of the week. When you finish an activity, note it on the chart and write down the time it was completed. Then determine its duration by subtracting the time the previous activity was finished from the newly entered time.

important academic obligations. If you have any long-range personal plans (concerts, weekend trips, etc.), enter the dates on the calendar as well. Keep your calendar up to date and refer to it often. Carry it with you at all times.

Develop a Weekly Calendar

Now that you have a general picture of the school term, develop a weekly schedule that includes all of your activities. Aim for a schedule that you can live with for the entire school term. A sample weekly schedule, which incorporates the following guidelines, is shown in Table 2.

1. Enter your class times, work hours, and any other fixed obligations first. *Be thorough.* Using information from your time-management diary, allow plenty of time for such things as commuting, meals, laundry, and the like.

2. Set up a study schedule for each of your courses. The study habits survey and your time-management diary will direct you. The following guidelines should also be useful.

   a. Establish regular study times for each course. The 4 hours needed to study one subject, for example, are most profitable when divided into shorter periods spaced over several days. If you cram your studying into one 4-hour block, what you attempt to learn in the third or fourth hour will interfere with what you studied in the first 2 hours. Newly acquired knowledge is like wet cement. It needs some time to "harden" to become fixed in your memory.

   b. Alternate subjects. The type of interference just mentioned is greatest between similar topics. Set up a schedule in which you spend time on several *different* courses during each study session. Besides reducing the potential for interference, alternating subjects will help to prevent mental fatigue with one topic.

   c. Set weekly goals to determine the amount of study time you need to do well in each course. This will depend on, among other things, the difficulty of your courses and the effectiveness of your methods. Many professors recommend studying at least 2–3 hours for each hour in class. If your time diary indicates that you presently study less time than that, do not plan to jump immediately to a much higher level. Increase study time from your baseline by setting weekly goals [see (4)] that will gradually bring you up to the desired level. As an initial schedule, for example, you might set aside an amount of study time for each course that matches class time.

   d. Schedule for maximum effectiveness. Tailor your schedule to meet the demands of each course. For the course that emphasizes lecture notes, schedule time for a daily review soon after the

TABLE 2 Sample Weekly Schedule

Time	Mon.	Tues.	Wed.	Thurs.	Fri.	Sat.
7–8	Dress Eat	Dress Eat	Dress Eat	Dress Eat	Dress Eat	
8–9	Psych.	Study Psych.	Psych.	Study Psych.	Psych.	Dress Eat
9–10	Eng.	Study Eng.	Eng.	Study Eng.	Eng.	Study Eng.
10–11	Study French	Free	Study French	Open Study	French	Study Stats.
11–12	French	Study Psych. Lab.	French	Open Study	French	Study Stats.
12–1	Lunch	Lunch	Lunch	Lunch	Lunch	Lunch
1–2	Stats.	Psych. Lab.	Stats.	Study or Free	Stats.	Free
2–3	Bio.	Psych. Lab.	Bio.	Free	Bio.	Free
3–4	Free	Psych.	Free	Free	Free	Free
4–5	Job	Job	Job	Job	Job	Free
5–6	Job	Job	Job	Job	Job	Free
6–7	Dinner	Dinner	Dinner	Dinner	Dinner	Dinner
7–8	Study Bio.	Study Bio.	Study Bio.	Study Bio.	Free	Free
8–9	Study Eng.	Study Stats.	Study Psych.	Open Study	Open Study	Free
9–10	Study	Study	Study	Open	Free	Free

This is a sample schedule for a student with a 16-credit load and a 10-hour-per-week part-time job. Using this chart as an illustration, make up a weekly schedule, following the guidelines outlined here.

class. This will give you a chance to revise your notes and clean up any hard-to-decipher shorthand while the material is still fresh in your mind. If you are evaluated for class participation (for example, in a language course), allow time for a review just *before* class meets. Schedule study time for your most difficult (or least motivating) courses during times when you are the most alert and distractions are fewest.

e. Schedule open study time. Emergencies, additional obligations, and the like could throw off your schedule. And you may simply need some extra time periodically for a project or for review in one of your courses. Schedule several hours each week for such purposes.

3. After you have budgeted time for studying, fill in slots for recreation, hobbies, relaxation, household errands and the like.

4. Set specific goals. Before each study session, make a list of specific goals. The simple note "7–8 pm. study psychology" is too broad to ensure the most effective use of the time. Formulate your daily goals according to what you know you must accomplish during the term. If you have course outlines with advance assignments, set systematic daily goals that will allow you, for example, to cover fifteen chapters before the exam. And be realistic. Can you actually expect to cover a 78-page chapter in one session? Divide large tasks into smaller units; stop at the most logical resting points. When you complete a specific goal, take a 5- or 10-minute break before tackling the next goal.

5. Evaluate how successful or unsuccessful your studying has been on a daily or weekly basis. Did you reach most of your goals? If so, reward yourself immediately.

You might even make a list of 5 to 10 rewards to choose from. If you have trouble studying regularly, you may be able to motivate yourself by making such rewards contingent on completing specific goals.

6. Finally, until you have lived with your schedule for several weeks, don't hesitate to revise it. You may need to allow more time for chemistry, for example, and less for some other course. If you are trying to study regularly for the first time and are feeling burned out, you probably have set your initial goals too high. Don't let failure cause you to despair and abandon the program altogether. Accept your limitations and revise your schedule so that you are studying only 15 to 20 minutes more each evening than you are used to. The point is to *identify a regular schedule with which you can achieve some success.* Time management, like any skill, must be practiced to become effective.

TECHNIQUES FOR EFFECTIVE STUDY

Knowing how to put study time to best use is, of course, as important as finding a place for it in your schedule. Here are some suggestions that should enable you to increase your reading comprehension and improve your note taking. A few study tips are included as well.

Using SQ3R to Increase Reading Comprehension

How do you study from a textbook? If you are like many students, you simply read and reread in a passive manner. Studies have shown, however, that most students who simply read a textbook cannot remember more than half the material ten minutes after they have finished. Often, what is retained is the unessential material rather than the important points upon which exam questions will be based.

This Study Guide is designed to facilitate, and allow you to assess, your comprehension of the important facts and concepts in the *Seasons of Life* telecourse. It will help you to integrate material from the audio and television programs with the content in Kathleen

Berger's text, *The Developing Person Through the Life Span, 2/e.* By learning the steps below you can improve your understanding of this, and any other, textbook. These steps make up a program known as SQ3R, which is an abbreviation for *Survey, Question, Read, Recite,* and *Review.*

Research has shown that students using SQ3R achieve significantly greater comprehension of textbooks than students reading in the more traditional passive manner.

Survey   Before reading a chapter, determine whether the text or the study guide has an outline or list of objectives. Read this material and the summary at the end of the chapter. Next, skim the textbook chapter and read the major headings and subheadings. This survey will give you an idea of the chapter's contents and organization. You will then be able to divide the chapter into logical sections in order to formulate specific goals for a more careful reading of the chapter.

Question   You will retain material longer when you have a use for it. If you look up a word's definition in order to solve a crossword puzzle, for example, you will remember it longer than if you merely fill in the letters as a result of putting other words in. Surveying the chapter will allow you to generate important questions for which the chapter will provide answers. These questions correspond to "mental files" into which knowledge will be sorted for easy access.

As you survey, jot down several questions for each chapter section. One simple technique is to generate questions by rephrasing a section heading. For example, the "Attachment" head could be turned into "What is Attachment?" Good questions will allow you to focus on the important points in the text. Examples of good questions are those that begin as follows: "List two examples of" "What is the function of?" "What is the significance of?" Such questions give a purpose to your reading. Alternatively, you may formulate questions based on the chapter outline.

Read   When you have established "files" for each section of the chapter, review your first question, begin reading, and continue until you have discovered its answer.

If you come to material that seems to answer an important question you don't have a file for, stop and write down the question.

Be sure to read everything. Don't skip photo or art captions, graphs, and marginal notes. In some cases, what may seem vague in reading will be made clear by a simple graph. Keep in mind that it is not uncommon for test questions to be drawn from this type of supplementary material.

Recite   When you have found the answer to a question, close your eyes and mentally recite the question and its answer. Then *write* the answer next to the question. It is important that you recite an answer in your own words rather than the author's. Don't rely on your short-term memory to repeat the author's words verbatim.

In responding to questions, pay close attention to what is called for. If you are asked to identify or list, do just that. If asked to compare, contrast, or do both, you should focus on the similarities (compare) and differences (contrast) between the concepts or theories. Answering questions carefully not only will help you to focus your attention on the important concepts of the text, but will also provide excellent practice for essay exams.

Recitation is an extremely effective study technique, recommended by many learning experts. In addition to increasing reading comprehension, it is useful for review. Trying to explain something in your own words clarifies your knowledge, often by revealing aspects of your answer that are vague or incomplete. If you repeatedly rely upon "I know" in recitation, you really *may not know.*

Recitation has the additional advantage of simulating an exam, especially an essay exam; the same skills are required in both cases. Too often students study without ever putting the book and notes aside, which makes it easy for them to develop false confidence in their knowledge. When the material is in front of you, you may be able to *recognize* an answer, but will you be able to *recall* it later, when you take an exam that does not provide these retrieval cues?

After you have recited and written your answer, continue with your next question. Read, recite, and so on.

Review   When you have answered the last question on the material you have designated as a study goal, go back and review. Read over each question and your written answer to it. Your review might also include a brief written summary that integrates all of your questions and answers. This review need not take longer than a few minutes, but it is important. It will help you retain the material longer and will greatly facilitate a final review of each chapter before the exam.

An excellent way to review your understanding of the chapters of *The Developing Person Through the Life Span, 2/e,* is to complete the "Key Questions" at the end of each textbook chapter and those in the "After Reading" section of this Study Guide. Then go through the "Lesson Guidelines" section of the Study Guide. You may be surprised to discover that you didn't know the chapter as well as you thought you did!

Also provided to facilitate your review are multiple-choice questions in the "Testing Yourself" section of the Study Guide. These questions cover both audio program and textbook content and should *not* be answered until you have listened to the program, read the chapter, and completed the "After Listening" and "After Reading" sections of the Study Guide. Correct answers, along with explanations of why each alternative is correct or incorrect, are provided at the end of the chapter. The relevant text page numbers for each question are also given. If you miss a question, read these explanations and, if you need to, review the text pages to further understand why. The questions do not test every aspect of a concept, so you should treat an incorrect answer as an indication that you need to review the concept.

One final suggestion. Incorporate SQ3R into your time-management calendar. Set specific goals for completing SQ3R with each assigned chapter. Keep a record of chapters completed and reward yourself for being so conscientious. Initially, it takes more time and effort to "read" using SQ3R, but with practice, the steps will become automatic. More important, you will comprehend significantly more material and retain knowledge longer than passive readers do.

Listening to the Audio Programs

Using audio tapes and television programs for learning requires much more active attention to their content than when these media are used simply for entertainment.

In following the steps outlined in each *Study Guide* lesson, you will gain the most from each program by applying the SQ3R method to audio tape listening. You will find it helpful to read the lesson orientation, lesson goals, and audio review questions before listening to the program. You may choose to listen to the entire program first and then answer the questions, replaying portions of the program as necessary. Or you may find it works better for you to answer the questions as you go along, stopping the tape as necessary. Soon after listening to the program, you should compare your answers with material in the "Lesson Guidelines" section of the Study Guide. Make sure that you have a good grasp of the answer to each audio question before you continue with the lesson.

Watching the Television Programs

The five television programs illustrate many of the facts and concepts of the course. They may be viewed any time during the semester and perhaps more than once. Before watching each program, you will find it helpful to read the program synopsis and story descriptions which appear in the Television Term Project chapter at the end of the Study Guide. To help focus your viewing on important series and program themes, you also should review the four essay questions that follow the story descriptions for each program.

Evaluating Your Exam Performance

How often have you received a grade on an exam that did not do justice to the effort you spent preparing for the exam? This is a common experience that can leave one feeling bewildered and abused. "What do I have to do to get an A?" "The test was unfair!" "I studied the wrong material!"

The chances of this happening are greatly reduced if you have an effective time-management schedule and use the study techniques described here. But it can happen, even to the best-prepared student. It is most likely to occur on your first exam with a new course.

Remember that there are two main reasons for studying. One is to learn for your own general academic development. Many people believe that such knowledge is all that really matters. Of course it is possible, though unlikely, to be an expert on a topic without achieving commensurate grades, just as one can, occasionally, earn an excellent grade without truly mastering the course material. During a job interview or in the work place, however, your A in Fortran won't mean much if you can't actually program a computer.

In order to keep career options open after you graduate, you must both know the material *and* maintain competitive grades. In the short run, this means performing well on exams, which is the second main objective in studying.

Probably the single best piece of advice to keep in mind when studying for exams is to *try to predict exam questions*. This means ignoring the trivia and focusing on the important questions and their answers (with your instructor's emphasis in mind).

A second point is obvious. How well you do on exams is determined by your mastery of *both* lecture (or, in this case, audio tape) and textbook material. Many students (partly because of poor time management) concentrate too much on one at the expense of the other.

To evaluate how well you are learning audio tape and textbook material, analyze the questions you missed on the first exam. Divide the questions into two categories, those drawn primarily from the audio tapes and those drawn primarily from the textbook. Determine the percentage of questions you missed in each category. If your errors are evenly distributed and you are satisfied with your grade, you have no problem. If you are weaker in one area, you will need to set future goals for increasing and/or improving your study of that area.

Similarly, note the percentage of test questions drawn from each category. While your instructor may not be entirely consistent in making up future exams, you may be able to tailor your studying by placing *additional* emphasis on the appropriate area.

Exam evaluation will also point out the types of questions your instructor prefers. Does the exam consist primarily of multiple-choice or essay questions? You may also discover that an instructor is fond of wording questions in certain ways. For example, an instructor may rely heavily on questions that require you to draw an analogy between a theory or concept and a real-world example. Evaluate both your instructor's style and how well you do with each format. Use this information to guide your future exam preparation.

Important aids, not only in studying for exams but also in determining how well prepared you are, are the "Testing Yourself" sections of the Study Guide. If these tests don't include all of the types of questions your instructor typically writes, make up your own practice exam questions. Spend extra time testing yourself with question formats that are most difficult for you. There is no better way to evaluate your preparation for an upcoming exam than by testing yourself under the conditions most likely to be in effect during the actual test.

A FEW PRACTICAL TIPS

Even the best intentions for studying sometimes fail. Some of these failures occur because students attempt to work under conditions that are simply not conducive to concentrated study. To help ensure the success of your self-management program, here are a few suggestions that should assist you in reducing the possibility of procrastination or distraction.

1. If you have set up a schedule for studying, make your roommate, family, and friends aware of this commitment, and ask them to honor your quiet study time. Close your door and post a "Do Not Disturb" sign.
2. Set up a place to study that minimizes potential distractions. Use a desk or table, not your bed or an extremely comfortable chair. Keep your desk and the walls around it free from clutter.
3. Do nothing but study in this place. It should become associated with studying so that it "triggers" this activity just as a mouth-watering aroma elicits an appetite.
4. Never study with the television on or with other distracting noises present. If you must have music in the background (in

order to mask dorm noises, for example) play soft instrumental music. Don't pick vocal selections; your mind will be drawn to the lyrics.

5. Study by yourself. Other students can be distracting or can break the pace at which *your* learning is most efficient. In addition, there is always the possibility that group studying will become nothing more than a social gathering. Reserve that for its own place in your time schedule.
6. Avoid studying in too many places. If you need a place other than your room, find one that meets as many of the above requirements as possible. Find a place, for example, in the library stacks.

If you continue to have difficulty concentrating for very long, try the following suggestions:

7. Study your most difficult or most challenging subjects first, when you are most alert.
8. Start with relatively short periods of concentrated study, with breaks in between. If your attention starts to wander, get up immediately and take a break. It is better to study effectively for 15 minutes and then take a break than to fritter away 45 minutes out of an hour. Gradually increase the length of study periods, using your attention span as an indicator of successful pacing.

SOME CLOSING THOUGHTS

We hope that these suggestions not only help make you more successful academically, but also enhance the quality of your college life in general. Not having the necessary skills makes any job a lot harder and more unpleasant than it has to be. Let us repeat the warning not to attempt to make too drastic a change in your lifestyle immediately. Start by establishing a few realistic goals, then gradually shape your performance to the desired level. Good habits require time and self-discipline to develop. Once established they can last a lifetime.

Introduction

AUDIO PROGRAM: Of Seasons, Stories, and Lives

ORIENTATION

Developmental psychology is concerned with how people change as they grow older and how they remain the same. According to the **life-span perspective,** development is a life-long process. Far from believing that our fates are sealed by the end of childhood, as some earlier theorists proposed, experts today believe that development continues throughout the life span and is unpredictable. The message is simple: You cannot tell how a life story will end just by knowing how it began.

Lesson 1 of the *Seasons of Life* series explores the many methods of life-span developmental psychologists. These include **naturalistic observation, case studies, interviews, experimentation,** as well as **longitudinal** and **cross-sectional** research. In Chapter 1 of *The Developing Person Through the Life Span, 2/e,* Kathleen Berger discusses the strengths and weaknesses of each method and two continuing controversies. These are the **nature–nurture issue** that addresses the relative importance of biological (nature) and environmental (nurture) influences on development, and the question of whether development is best viewed as a **continuous process** or a sequence of distinct **stages.**

Audio program 1, "Of Seasons, Stories, and Lives," notes that one method of developmental research is the interpretation of **life stories.** We hear several people recall their earliest memories and their memories of **nuclear episodes:** the most significant moments in their life stories. Through the expert commentary of psychologist Dan McAdams, we learn what such **autobiographical memories** tell researchers about a person's life story. With the help of commentary by psychologist Richard Lerner, we come to a fuller appreciation of the life-span perspective.

Lesson 1 also introduces three **developmental clocks** that govern our progress through the seasons of life. The **biological clock** is a metaphor for the body's way of timing its physical development. The **social clock** reflects society's age norms for when certain life events should occur. The **psychological clock** represents each person's inner timetable for development.

As the program opens, we hear a woman recounting her earliest memory—the opening scene of her life story.

LESSON GOALS

By the end of this lesson you should be prepared to:

1. Define developmental psychology and explain the life span and the ecological perspectives.

2. Describe the various research methods used by developmental psychologists, noting the strengths and weaknesses of each.

3. Discuss the significance of the three developmental clocks through the life span.

4. Explain how the interpretation of life stories and autobiographical memories helps psychologists to understand development through the life span.

Audio Assignment

Listen to the audio tape that accompanies Lesson 1: "Of Seasons, Stories, and Lives."

Write answers to the following questions. You may replay portions of the program if you need to refresh your memory. Answer guidelines may be found in the Lesson Guidelines section at the end of this chapter.

1. Explain the "life-span perspective" in developmental psychology and the concept that there are "seasons" of life.

can't understand young or out old or anyone in between w/out understanding both young & old.

2. Compare and contrast the three developmental clocks.

Biological - Body keeping time - changing shapes a person for society
Social - Societies
reveals self→Psychol - Individuals way - How we feel

3. Explain the significance of first memories and nuclear episodes in the interpretation of life stories.

Autobiographical memory begins — people choose their first memories — begins one's current identity.

significant scene best or worst

how you describe nuc episodes reveals personality traits, reveals who we really are - ordinary episodes don't

Textbook Assignment

Read Chapter 1: "Introduction," pages 1–29 in *The Developing Person Through the Life Span, 2/e.*

Write your answers to the following questions. Refer back to the textbook, if necessary. Answer guidelines may be found in the Lesson Guidelines section at the end of this chapter.

1. Define developmental psychology.

    How and why people change over time and how and why they stay the same

2. Identify and define the three domains used in the study of human development.

    1) physical — body changes, motor skills

    2) cognitive — intellect, thought, language

    3) psychosocial — emotions, personality, relationships

3. Identify new directions in developmental psychology and explain the ecological, or systems, approach to the study of human development.

    Broader focus — more aware of how the individual is affected by many other individuals and groups — also how history, culture politics, economics are shaping the individual.

    ecological or systems approach — emphasizes the impact of society, culture and physical setting — as well as other people's affect,

4. Identify and discuss two major areas of controversy in developmental psychology today.

    Nature — inherited traits

    Nurture — environmental influences

    Continuity — gradual changes

    Discontinuity — rapid & radical change

5. Cite the steps usually involved in applying the scientific method.

    1 - Form a research question
    2 - Develop a hypothesis
    3 - Test hypothesis
    4 - Draw conclusions
    5 - Make findings available

6. Describe the major research methods used by developmental psychologists and cite the strengths and weaknesses of each method.

    — Naturalistic observation — does not pinpoint cause and effect
    — Laboratory experiment — pinpoints but does not relate to daily life
    — Interview — interviewer may give answers that interviewer wants to hear — biased interviewer — mood of interviewee
    — Case study — not broad enough —

7. Compare and contrast longitudinal, cross-sectional, and sequential research methods.

    more difficult — longitudinal — studies individuals over a long period of time

    cross sectional — compares people of different ages but similar in other important ways

    sequential — combines both — many different people tested over a long period of time — thw a new group of people tested at each age level

Testing Yourself

After you have completed the audio and text review questions, see how well you do on the following quiz. Correct answers, with text and audio references, may be found at the end of this chapter.

1. Which of the following statements most accurately expresses the life-span perspective on development?
   a. Human development reflects the interaction of three developmental clocks.
   b. Human development is a continuous process rather than a series of stages.
   c. Human development occurs in discontinuous stages.
   d. Human development is a lifelong process of change.

2. The research method in which the same group of people is studied over a long period of time is called the:
   a. cross-sectional method.
   b. longitudinal method.
   c. sequential method.
   d. naturalistic method.

3. Culturally determined age norms for when to enter school, when to start a family, and when to retire, are set according to:
   a. the biological clock.
   b. the social clock.
   c. the psychological clock.
   d. all three developmental clocks.

4. Memories of especially significant events in a life story are called:
   a. flashbulb memories.
   b. iconic memories.
   c. nuclear episodes.
   d. eidetic images.

5. Dan McAdams studies autobiographical memories by:
   a. coding them for the presence of certain motives.
   b. interviewing relatives of the subject of a case study.
   c. using the cross-sectional method.
   d. interpreting art produced by subjects at different stages in their life span.

6. The ecological, or systems, approach to the study of development stresses the:
   a. context in which development occurs.
   b. cognitive domain.
   c. physical domain.
   d. discontinuity of development.

7. A major advantage of the experimental method is that:
   a. its results are usually applicable to daily life.
   b. it may uncover cause-and-effect relationships.

    c. it eliminates the need for statistical analysis.

    d. all of the above are true.

8. Testable predictions regarding behavior that form the basis of an experiment are called:

    a. theories.

    b. operational definitions.

    c. premises.

    d. hypotheses.

9. The psychosocial domain is largely concerned with:

    a. intellectual development.

    b. personality, emotions, and relationships.

    c. maturation.

    d. identifying stages of development.

10. Stage theories of human development emphasize that:

    a. development is based on the interaction of several "systems."

    b. development is continuous.

    c. development is discontinuous.

    d. development is unpredictable.

11. If two variables are strongly related, their correlation is:

    a. positive.

    b. negative.

    c. close to 0.00.

    d. close to 1.0.

12. A psychologist who watches the behaviors of people in their usual surroundings is engaged in which type of research?

    a. Correlational

    b. Experimental

    c. Naturalistic observation

    d. Case study

13. The control group of an experiment is one in which the:

    a. subjects are randomly assigned to various conditions.

    b. experimental treatment of interest is absent.

    c. subjects receive the experimental treatment of interest.

    d. experimenter is unaware of which conditions are in effect.

14. In the ecological approach, the mesosystem refers to:

    a. immediate systems, such as the family, that affect the individual.

    b. the interlocking systems that link one microsystem to another.

    c. the neighborhood and community structures that affect the functioning of smaller systems.

    d. the overarching patterns of culture, politics, and the economy.

15. Nature is to nurture as:

    a. genetic influence is to environmental influence.

    b. environmental influence is to genetic influence.

    c. microsystem is to mesosystem.

    d. mesosystem is to microsystem.

LESSON 1 EXERCISE: FIRST MEMORIES

The audio program notes that one way of understanding an individual's life is by listening to his or her life story. In order to make sense of life stories, psychologists are beginning to probe **autobiographical memory.** This kind of memory begins with each individual's very first memory of life. Although first memories are often a mixture of fact and fiction, they are especially revealing glimpses into each person's current identity.

To further your understanding of the material presented in this lesson, the Exercise asks you to reflect on the first memories of characters heard in the audio program. Answer the following questions and return your completed **Exercise Response Sheet** to your instructor.

LESSON GUIDELINES

Audio Question Guidelines

1. The life-span perspective states that development is not fixed early in the life span, as earlier theorists had proposed, but continues throughout the seasons of life.

   Developmental psychologists recognize separate life-span seasons of infancy, childhood, adolescence, early adulthood, middle adulthood, and late adulthood.

   The life-span perspective also emphasizes that people become more diverse as they age, that they are capable of controlling their development, that is, of being the authors of their own life stories.

   The life-span perspective has also made experts aware of the three developmental clocks—the biological clock, social clock, and psychological clock—that interrelate to pace human development.

2. The developmental clocks are metaphors for the three ways in which people change during the seasons of life. One reason people change is because their bodies change. The **biological clock** represents the body's mechanisms of timing physical development. It times birth, growth, the reproductive cycle, and aging.

   People also change because the world around them changes. The **social clock** represents each society's age norms for when certain life events, such as entering school, starting a family, and retiring, should occur.

   A third reason people change is because their own inner needs change. The **psychological clock** represents a person's inner timetable for development, for his or her own way of determining the right time for certain life events.

3. One way of understanding life-span development is to listen to a specific person's life story. The story, although continually evolving and being rewritten, may be what gives each life continuity and purpose. It may represent each individual's sense of who he or she is.

   The interpretation of **autobiographical memory**, including **nuclear episodes** and first memories, is especially interesting. Cognitive psychologists suggest that what we remember from the past is reconstructed over time so that such memories are generally combinations of fact and fantasy.

   Early memories, whether true or not, say something about a person's current identity and highlight significant themes in his or her entire life story.

   Nuclear episodes are personal memories that are very significant to the individual. How people describe nuclear episodes often reveals important personality traits, for example, a desire for intimacy.

Textbook Question Guidelines

1. The primary goal of developmental psychology is to help every person achieve his or her fullest potential.

   The study of human development is the study of how and why people change over time, as well as the study of how and why they remain the same.

2. The **physical domain** includes body changes and motor skills.

   The **cognitive domain** includes intellect, thought processes, and language.

   The **psychosocial domain** includes emotions, personality, and relationships with other people.

   Development is **holistic,** with development in one domain affecting development in all the other domains.

3. Formerly focused on the individual, developmental psychology has taken a broader focus in recent years, to become more aware of how the individual interacts with others and is influenced by external social forces such as history, culture, politics, and economics.

   According to Urie Bronfenbrenner, the systems that support human development occur at four nested levels: the **microsystem** (immediate setting), **mesosystem** (relationship between various microsystems), the **exosystem** (major social structures), and the **macrosystem** (overarching patterns of culture).

4. The relative importance of **nature** (inherited capacities) and **nurture** (environmental influences) remains controversial, because it is difficult to prove which is more responsible for a particular developmental change. The text takes the position that the two influences interact in most aspects of development, rather than conflicting or functioning separately.

   The **continuity–discontinuity** debate concerns whether development is a gradual growth process (continuity) or occurs in stages (discontinuity). The discontinuity view of development has dominated developmental theory in the twentieth century.

   To a certain degree, however, developmental stages are cultural creations. Emphasis on stages also may miss many differences, inconsistencies, and irregularities that are important features in development.

5. The five basic steps of the scientific method are (1) formulating a research question, (2) developing a hypothesis, (3) testing the hypothesis, (4) drawing conclusions, and (5) reporting the results.

   The ability to reproduce, or **replicate,** research findings is of vital importance to the scientific method.

   Scientists attempt to increase the validity of their results by using a **representative sample,** separate **experimental** and **control groups,** and "**blind**" **experimenters.**

6. **Naturalistic observation** is a research method in which people are observed in their usual surroundings (home, school, work place).

   In the **case study,** the life history of one individual is studied in depth.

   The **interview** is a research method in which a scientist asks a group of people specific questions to discover their opinions on a particular topic.

   In an **experiment,** people are brought to a controlled setting in order to allow a scientist to manipulate a specific variable and observe its effects on their behavior. Because variables are directly controlled in an experiment, it may yield information regarding cause-and-effect relationships in development.

Because the variables in naturalistic observation are numerous and uncontrolled, it is difficult to pinpoint which of them might be the cause of a particular event.

The interview is subject to several types of bias, including the fact that the phrasing of questions may influence answers, that people may give the answers they think the interviewer wants to hear, and that the interpretation of answers may depend on the interviewer's bias.

One limitation of the experiment is that subjects may act differently than they would normally. A second weakness of this method is that even if subjects behave candidly in the experiment, their performance may not generalize to similar situations in the real world.

7. In **cross-sectional** research, groups of people who are different in age but similar in all other important characteristics are compared on the characteristic of interest.

One limitation of cross-sectional research is that it is always possible that some variable other than age differentiates the groups.

In **longitudinal research** the sample group of people is studied over a period of time.

Longitudinal research is particularly useful in studying developmental trends that occur over a long period.

Both longitudinal and cross-sectional research are subject to **cohort effects,** according to which a group of people born in a particular generation may experience unique social conditions.

Sequential research, based on a combination of cross-sectional and longitudinal techniques, is designed to control for such cohort effects.

Answers to Testing Yourself

1. **d.** In contrast to earlier views that development was largely fixed early in life, the life-span perspective views development as continuing throughout the seasons of life. (audio program)

2. **b.** The *longi*tudinal method studies development over a *long* period of time. (audio program)

3. **b.** The social clock represents society's way of telling us to "act our age." (audio program)

4. **c.** The nucleus is "at the center." Nuclear episodes are central to our sense of who we are. (audio program)

5. **a.** In McAdams' view, life stories reflect each person's sense of identity. (audio program)

6. **a.** The ecological approach emphasizes the interaction between and among various systems that influence development. (textbook, pp. 5–6)

7. **b.** Because variables are tested in a controlled environment, the link between cause and effect is much clearer in an experiment than in other research methods. (textbook, p. 22)

8. **d.** Formulating a hypothesis is an integral phase in the scientific method. (textbook, p. 16)

9. **b.** Answer "a" reflects the cognitive domain, "c" the physical domain, and "d" has no bearing on the three domains. (textbook, p. 4)

10. **c.** Discontinuity theorists see development as best described by age-related periods of rapid and radical change. (textbook, p. 13)

11. **d.** Answers "a" and "b" are incorrect because positive and negative correlations may be equally strong. A correlation of zero indicates no relationship between two variables. (textbook, p. 21)

12. **c.** Naturalistic observation has the advantage that it involves studying people in their everyday environments. (textbook, p. 17)

13. **b.** The control group serves as a point of comparison for the experimental group in an experiment. (textbook, p. 19)

14. **b.** Answer "a" describes the microsystem, "c" the exosystem, and "d" the macrosystem. (textbook, pp. 5–6)

15. **a.** "Nature," indicates an inherited capacity; "nurture" indicates the influence of the environment. (textbook, p. 11)

References

Neugarten, B. L. (1985). Interpretive social science and research on aging. In Alice S. Rossi (ed.) *Gender and the life course,* New York: Aldine Publishing Co., pp. 291–300.

Professor Neugarten, an eminent sociologist who provides expert commentary in the *Seasons of Life* television series, makes a persuasive argument that the research methods of the natural sciences are of limited use in the study of life-span development.

Rubin, D. C. (1985). The subtle deceiver: recalling our past. *Psychology Today,* (September), pp. 39–46.

Professor Rubin discusses research on autobiographical memory and the insights it yields into each person's sense of him or herself.

Theories

AUDIO PROGRAM: The Story of Erik Erikson

ORIENTATION

Lesson 1 introduced the subject matter of developmental psychology, described the three clocks that govern development, and explained the various research methods used in life-span psychology. Lesson 2 deals with theories of human development, including that of Erik Erikson, who is the subject of the audio program.

Theories provide a useful way of organizing ideas about behavior into testable **hypotheses.** In Chapter 2 of *The Developing Person Through The Life Span, 2/e,* four theories that have significantly influenced life-span psychology are compared and evaluated. The theories complement one another: each emphasizes different aspects of development and is too restricted to be used on its own to explain the diverse ways in which development occurs. **Psychoanalytic theory** focuses on early experiences and distinguishable stages of growth, **learning theory** on environmental influences. **Cognitive theory** emphasizes the influence of thinking on behavior, and **humanistic theory** stresses the positive aspects of growth throughout life.

Most developmental psychologists today take an **eclectic perspective,** applying insights from various theories rather than limiting themselves to only one school of thought. As the author of the text points out, the final test of a theory is its usefulness in clarifying observations and suggesting new hypotheses.

One theory that has withstood this test is that of Erik Erikson. A student of Freud, Erikson was one of the first psychologists to devote attention to the entire life cycle—to adulthood as well as to childhood. Erikson spent his childhood in Germany and came to America when Adolf Hitler became Chancellor and Freud's writings were publicly burned in Berlin. In America, Erikson practiced as one of the first psychoanalysts for children, studying people as diverse as the Native American Sioux and soldiers who suffered emotional trauma during World War II. Erikson's experiences led him to the conclusion that Freud's **psychosexual** stages were too limited. In Erikson's view there were eight important **crises** in life and, hence, eight (rather than five) stages of development. Unlike Freud's stages, Erikson's reflect social and cultural influences; as a result, they are called **psychosocial stages.** In the program, these stages are outlined and contrasted with those embodied in Freud's theory.

The program begins with Erikson himself describing how he came to create one of psychology's most influential theories of the life cycle.

LESSON GOALS

By the end of this lesson you should be prepared to:

1. Explain the role theories play in developmental psychology.

2. Outline the basic terms and themes of the four major theories of human development: psychoanalytic theory, learning theory, humanistic theory, and cognitive theory.

3. Discuss the eclectic perspective in developmental psychology.

4. Describe Erikson's eight stages of psychosocial development, and discuss the significance of Erikson's theory in life-span psychology.

Audio Assignment

Listen to the audio tape that accompanies Lesson 2: "The Story of Erik Erikson."

Write answers to the following questions. You may replay portions of the program if you need to refresh your memory. Answer guidelines may be found in the Lesson Guidelines section at the end of this chapter.

1. Outline Freud's five stages of psychosexual development.

2. Outline Erikson's eight stages of psychosocial development. Explain how his theory diverges from that of Freud, and discuss its significance in life-span psychology.

3. Cite several contributions Erikson's psychosocial theory has made to the study of development, and several criticisms of it.

Textbook Assignment

Read Chapter 2: "Theories," pages 31–53 in *The Developing Person Through the Life Span, 2/e.*

Write your answers to the following questions. Refer back to the textbook, if necessary. Answer guidelines may be found in the Lesson Guidelines section at the end of this chapter.

1. Describe the use of theories in scientific investigation.

2. Describe Freud's three components of personality. Cite several contributions psychoanalytic theory has made to the study of development, and several criticisms of it.

3. Describe the learning theory approach to understanding human development.

4. Cite several contributions learning theory has made to the study of development, and several criticisms of it.

5. Contrast the theories of Maslow and Rogers with those emphasizing psycho-analysis and behaviorism, and cite several contributions and criticisms of humanistic theory.

6. Outline Piaget's four stages of cognitive development and his theory of the processes by which mental growth and adaptation occur.

7. Cite several contributions cognitive theory has made to the study of development, and several criticisms of it.

8. Compare and contrast the four major theories of development and explain the eclectic perspective.

Testing Yourself

After you have completed the audio and text review questions, see how well you do on the following quiz. Correct answers, with text and audio references, may be found at the end of this chapter.

1. Erik Erikson's theory of development:
   a. is based on eight crises all people are thought to face.
   b. emphasizes cultural and social influences on development.
   c. was one of the first to emphasize that development was life-long.
   d. includes all of the factors listed above.

2. Freud's stages of development are called _____ stages; Erikson's are called _____ stages.
   a. psychosexual . . . psychosocial
   b. psychosocial . . . psychosexual
   c. psychoanalytic . . . social learning
   d. psychoanalytic . . . neo-Freudian

3. According to Freud's theory, the correct sequence of stages of development is:
   a. oral, anal, genital, latent, phallic.
   b. anal, oral, phallic, latent, genital.
   c. genital, oral, anal, latent, phallic.
   d. oral, anal, phallic, latent, genital.

4. In Erikson's theory, infants experience the crisis of:
   a. autonomy vs. shame.
   b. trust vs. mistrust.
   c. industry vs. role confusion.
   d. identity vs. role confusion.

5. During the period from ages 6 to 12, when Freud considered that sexual forces become dormant, Erikson saw a critical time of conflict between:
   a. autonomy and shame.
   b. trust and mistrust.
   c. industry and inferiority.
   d. identity and role confusion.

6. According to the textbook, theories are:
   a. testable predictions.
   b. collections of general ideas expressed in a clear framework.
   c. hunches about cause-and-effect relationships.
   d. all of the above.

7. Psychoanalytic theory emphasizes that human development is influenced by:
   a. thinking about our experiences.
   b. environmental influences.
   c. unconscious drives.
   d. our natural growth-promoting tendencies.

8. One criticism of psychoanalytic theory is that it:
   a. offers too simple an explanation of human development.
   b. is subjective and difficult to test.
   c. fails to take into consideration that humans act irrationally.
   d. is too psychological and fails to consider our biological nature.

9. According to learning theory, a child who behaves very aggressively is doing so because he or she:
   a. is displacing unconscious impulses.
   b. is obviously frustrated.
   c. has been reinforced for acting this way.
   d. has an aggressive nature.

10. Humanistic theorists such as Rogers and Maslow have criticized psychoanalytic theory for its emphasis on:
    a. unconscious impulses, rather than observable behaviors.
    b. the ways in which people are different, rather than their similarities.
    c. psychological illness, rather than healthy development.
    d. all of the above.

11. Cognitive theories emphasize:
    a. biological forces in development.
    b. individual differences in development.
    c. how people's expectations influence their behavior.
    d. observable behaviors.

12. According to Piaget's theory, development occurs in the following sequence of stages:
    a. Sensorimotor, preoperational, concrete operational, formal operational
    b. Preoperational, sensorimotor, concrete operational
    c. Concrete operational, sensorimotor, preoperational, formal operational
    d. Preoperational, concrete operational, sensorimotor, formal operational

13. A developmental psychologist who takes an eclectic perspective:
    a. views human behavior as too unpredictable to be studied scientifically.
    b. emphasizes external, rather than internal, influences on behavior.
    c. does not subscribe to any one theory of behavior.
    d. sees development as a continuous process, rather than occurring in stages.

14. Which theory proposes that, rather than occurring in stages, development is a continuous process based on the same principles throughout the life span?
    a. Learning theory
    b. Psychosocial theory
    c. Psychosexual theory
    d. Piaget's theory

15. In comparing the four major theories of human development, the author of *The Developing Person Through The Life Span* concludes that:
    a. once definitive experiments are conducted, one of the theories will prove correct.
    b. each is flawed in being based on the same underlying premise: that development is predictable.
    c. each is too "mechanistic" to account for the complexities of human development.
    d. each has contributed a great deal to the study of human development.

LESSON 2 EXERCISE: THEORIES OF HUMAN DEVELOPMENT

Four major theories of human development are described, compared, and evaluated in Chapter 2. These are the **psychoanalytic theories** of Freud and neo-Freudians such as Erikson; the **learning** and **social learning theories** of Pavlov, Skinner, and Bandura; the **humanistic theories** of Maslow and Rogers; and the **cognitive theory** of Piaget. Although each theory is too restricted to account solely for the tremendous diversity in human development, each has made an important contribution to life-span psychology.

To help clarify your understanding of the major developmental theories, this exercise asks you to focus on the similar, contradictory, and complementary aspects of the four theories. Please write your answers to the following questions on the **Exercise Response Sheet.** Return your completed response sheet to your instructor.

LESSON GUIDELINES

Audio Question Guidelines

1. According to Freud's theory of infantile sexuality, children have sexual pleasures and fantasies long before adolescence.

   During the first five or six years, development progresses through three **psychosexual stages,** characterized by the focusing of sexual interest and pleasure, successively, on the mouth **(oral stage),** the anus **(anal stage),** and the sexual organs **(phallic stage).**

   Freud believed that personality was well established by the end of stage three, about the age of 6.

   Following a five- or six-year period of sexual **latency,** the individual enters the **genital stage,** which lasts throughout adulthood.

2. Erikson's **psychosocial stage** theory emphasizes each person's relationship to the social environment. Erikson proposed eight developmental stages, each characterized by a particular crisis that must be resolved in order for the individual to progress developmentally.

   During the first stage **(trust vs. mistrust),** babies learn either to trust or to mistrust that others will meet their basic needs.

   During the second and third years of life **(autonomy vs. shame and doubt)** children learn either to be self-sufficient in many activities or to doubt their own abilities.

   During the third stage **(initiative vs. guilt)** children begin to envisage goals and to undertake many adultlike activities, sometimes experiencing guilt as they overstep the limits set by their parents.

   During the years from 6 to 12 **(industry vs. inferiority)** children busily learn to feel useful and productive; failing that, they feel inferior and unable to do anything well.

   At adolescence **(identity vs. role confusion)** individuals establish sexual, ethnic, and career identities or become confused about who they are.

   Young adults **(intimacy vs. isolation)** seek companionship and love from another person or become isolated from others.

   Middle-aged adults **(generativity vs. stagnation)** feel productive in their work and family or become stagnant and self-absorbed.

   Older adults **(integrity vs. despair)** try to make sense out of their lives; they either see life as meaningful or despair in their failure to attain goals.

3. Erikson was one of the first psychologists to view development as a life-long process, not one largely fixed by the end of childhood, as Freud had proposed.

   Erikson also emphasized the importance of cultural and social influences on development.

   Some critics say that Erikson's theory is biased toward male development.

The principal objection to Erikson's theory is its basic outline of life as a sequence of fixed stages. Critics argue that development is much more variable and flexible than this discontinuous, stage approach implies.

Textbook Question Guidelines

1. **Theories** provide a framework for ideas that permit a cohesive view of the complexities that may be involved in any human behavior.

   Theories can be used to organize a researcher's assumptions into **hypotheses** that can be tested.

   The value of a theory can be measured by how productive it is in generating hypotheses to test and in inspiring new insights into behavior.

2. According to Freud, the three components of personality are the **id, ego,** and **superego.**

   Operating according to the **pleasure principle,** the id is the source of unconscious impulses toward the fulfillment of needs.

   As the infant begins to learn that gratification of its own needs must sometimes wait, the **ego** develops. Operating according to the **reality principle,** the ego attempts to satisfy the demands of the id in ways that accommodate to the real world.

   At about age 4 or 5, the **superego** starts to develop, as children begin to identify with their parents' moral standards.

   Developmentalists have been influenced by the psychoanalytic insights that (1) at different stages of life a person has different problems and needs, (2) certain periods of the life span are particularly important in development, and (3) human actions are likely to be far more complex than might initially be apparent.

   Most researchers agree that personality characteristics are affected more by genetic traits, current life events, and sociocultural context than by the experiences of early childhood, as Freud theorized.

   Freud's theory is also criticized as being an anachronism of 19th century morality.

   Psychoanalytic theory is also criticized as being untestable.

3. Instead of developing a stage theory, **learning theorists** (frequently called **behaviorists**) have formulated laws of behavior that are believed to operate at every age.

   According to this view, life is a continual process of **conditioning,** in which new behaviors are responses to old stimuli and unproductive responses are eliminated.

   Classical conditioning involves learning by association: the individual learns to associate a neutral stimulus with a meaningful one.

   According to **operant conditioning,** the individual's past history of **reinforcement** and **punishment** influences his or her future behavior.

   Social learning theory emphasizes the ways in which people learn by observing others **(modeling)**, in addition to learning from direct experience and reinforcement.

4. Learning theory's emphasis on the causes and consequences of specific behaviors has led to the realization that many seemingly inborn behavior patterns may actually be the result of interaction with the environment. This realization has prompted a new, more pragmatic approach to the treatment of many problem behaviors.

   The scientific rigor of learning theory provides a model for developmentalists that encourages them to define terms more precisely, test hypotheses, and generally follow a more objective methodology.

   Learning theorists have been criticized for limiting their understanding of behavior by ignoring human emotions, ideas, the unconscious, and any unobservable behaviors.

   Because learning theorists' formulation of general laws of behavior was primarily based on research with lower animals, learning theory may have limited relevance to human behavior.

5. **Humanistic theory** takes issue with psychoanalytic theory's emphasis on abnormal behavior and the mechanistic orientation of behaviorists.

   The **holistic** view embodied in humanistic theory considers a person to be a whole, unique being, rather than a collection of drives, instincts, and stimulus–response relationships.

   At the core of each person is a drive to realize his or her potential and achieve **self-actualization.**

   Humanistic theorists emphasize that growth and development of the **self** can occur throughout the life span.

   The broad vision of the humanists balances the narrow views of some behaviorists.

   Maslow proposed a **hierarchy of needs,** with physiological needs at the base and self-actualization at the apex, and argues that if a person's lower needs are not met, he or she cannot gratify loftier needs.

   Rogers applied humanistic theory to the therapeutic setting, believing that each person is driven to become a **fully functioning** human being.

   Humanists have been criticized for overlooking the many ways in which society, families, and individuals may prevent the full development of an individual's human potential.

6. Piaget's periods of cognitive development begin with the sensorimotor period (birth–2 years), during which the infant uses sensory and motor abilities to understand the world. During the **preoperational** period (2–6 years) the child is able to use **symbolic thinking,** including language, to understand the world. Preschoolers' thinking is **egocentric,** however. During the **concrete operational** period (7–11 years), the child understands and applies logical operations to help interpret specific experiences. From age 12 on, in the **formal operational** period, the individual is able to think about abstractions and hypothetical concepts.

According to Piaget, each person strives for **equilibrium** between existing mental concepts (**schemas**) and new experiences. Intelligence involves the organization of ideas and adaptation of existing schemas to new perceptions.

Adaptation can occur in two ways, through **assimilation** (new information is fitted into existing schemas), and **accommodation** (schemas are altered to take new information into account).

7. Cognitive theory has revolutionized developmental psychology by focusing attention on active mental processes. It has also helped researchers in many fields become aware of their own subjectivity and how that influences their observations.

   Some critics consider that Piaget underestimated the role of external motivation, teaching, society, and the home in fostering a child's cognitive development.

   Because it focuses primarily on scientific and logical thinking, Piaget's description of intelligence may have limited applicability in daily life.

   Critics also argue that Piaget's stages may not be universally applicable.

8. The four theories complement one another: each emphasizes somewhat different aspects of development.

   Psychoanalytic theory, although subjective, has drawn attention to the importance of early experiences in development.

   Learning theory, although mechanistic, has indicated the effect that the environment has on behavior.

   Cognitive theory has led to a greater understanding of how our thinking affects our actions.

   Humanistic theory has emphasized life-long growth and a more positive developmental perspective.

   Today, most developmental psychologists have an **eclectic** perspective; they apply insights from various theories rather than limiting themselves to adherence to one school of thought.

Answers to Testing Yourself

1. **d.** All of these are true of Erikson's theory. (audio program)

2. **a.** Freud's stages focus on gratification of sexual pleasure; Erikson's stages focus on each person's relationship with the social environment. (audio program; textbook, p. 36)

3. **b.** (audio program; textbook, p. 37)

4. **b.** Infants learn either to trust or mistrust that others will meet their basic needs. (audio program, textbook, p. 37)

5. **c.** During this stage, according to Erikson, children learn to be competent and productive or feel inferior and unable to do anything well. (audio program, textbook, p. 37)

6. **b.** Theories provide a cohesive framework in which to collect ideas. (textbook, p. 31)

7. **c.** Psychoanalytic theory interprets human development in terms of unconscious drives and motives. (textbook, p. 32)

8. **b.** Freud's theory offers few, if any, hypotheses that are testable. (textbook, p. 52)

9. **c.** Learning theory proposes that behaviors such as aggression are learned and maintained through reinforcement. (textbook, p. 40)

10. **c.** Humanistic theorists take issue with what they see as psychoanalytic theory's emphasis on abnormal behavior. (textbook, p. 44)

11. **c.** The prime focus of cognitive theory is the way in which the individual's thought processes and expectations influence his or her behavior. (textbook, p. 47)

12. **a.** (textbook, p. 48)

13. **c.** Rather than adopting any one theory exclusively, eclectic theorists make use of all of them. (textbook, p. 52)

14. **a.** Unlike adherents of the other theories discussed, learning theorists believe that development occurs according to the same principles at every age. (textbook, p. 38)

15. **d.** Although each theory is too restricted to account for the diversity of human development, each has made us aware of some important aspects of behavior. (textbook, p. 52)

References

Erikson, Erik H. (1963). *Childhood and society* (2nd ed.) New York: Norton.

This is Erikson's landmark publication, which outlines the eight stages of psychosocial development.

Genetics

AUDIO PROGRAM: And Then We Knew: The Impact of Genetic Information

---

ORIENTATION

Lesson 1 and 2 of *Seasons of Life* examined the meaning of life stories and the methods and theories of life-span development. Now we turn to the journey through life itself. Lesson 3 focuses on **genetics,** the science concerned with the mechanisms of biological inheritance. The audio program and text explain how physical characteristics are inherited from our parents through **genes, chromosomes,** and **DNA.** They also describe **chromosomal abnormalities,** which occur when a fertilized egg has too few or too many chromosomes, and the physical and mental disorders that may be the result.

As described in Chapter 3 of the text, and by the experts in the audio program, **genetic testing** can help to predict whether a couple will produce a child with a genetic problem. In addition, the emerging field of **genetic counseling** plays a vital role in helping people to understand and cope with genetic information. Through the new experimental techniques of **gene mapping** and **gene replacement therapy,** researchers are gaining a deeper understanding of the causes of many genetic disorders and of how to prevent them.

As the audio program illustrates, the price of advances in genetic technology is increased knowledge, the implications of which many individuals would rather not confront. Knowing that the husband is a carrier of a deleterious gene, a young couple faces the difficult decision of whether or not to have children. In addition to its potentially devastating impact on Don's self-esteem, the genetic "news" deeply affects Karen, and Don's mother. As the story unfolds, we learn from geneticist Dr. Donald Rucknagel and genetic counselor Diane Baker of the incredible technological advances that have made genetic counseling possible, and of the impact this technology has on a real couple and their extended families.

LESSON GOALS

By the end of this lesson, you should be prepared to:

1. Explain the basic mechanisms of heredity, including the significance of chromosomes and genes.

2. Describe common causes of genetic abnormalities and several techniques of genetic testing for the presence of such disorders.

3. Discuss the process and importance of genetic counseling.

Audio Assignment

Listen to the audio tape that accompanies Lesson 3: "And Then We Knew."
Write answers to the following questions. You may replay portions of the
program if you need to refresh your memory. Answer guidelines may be found
in the Lesson Guidelines section at the end of this chapter.

1. What is the difference between a person's **genotype** and **phenotype**?

2. In the audio program, how was **amniocentesis** used to determine that Karen
and Don's first child would have been mentally retarded and physically
deformed?

3. What is the baseline genetic risk factor that is present in any pregnancy?

4. How can genetic counseling help in each of the following areas?

   a. prenatal diagnosis

   b. pediatric genetics

   c. adult-onset conditions

5. What is the significance of each of the following techniques for treating
genetic abnormalities?

   a. chorionic villus sampling

   b. gene mapping

   c. gene replacement therapy

Textbook Assignment

Read Chapter 3: "Genetics," pages 55–75 in *The Developing Person Through The Life Span, 2/e.*

Write your answers to the following questions. Refer back to the textbook, if necessary. Answer guidelines may be found in the Lesson Guidelines section at the end of this chapter.

1. What is the difference between monozygotic and dizygotic twins? How have twins been used in developmental research?

2. Why are most human characteristics considered to be multifactorial in nature?

3. Which chromosomal abnormalities are identified in the textbook? Describe their effects and their causes.

4. What are some of the causes of genetic problems?

Testing Yourself

After you have completed the audio and text review questions, see how well you do on the following quiz. Correct answers, with text and audio references, may be found at the end of this chapter.

1. Genes, arranged along the DNA molecule, along with other materials, make up a:
   a. zygote.
   b. chromosome.
   c. specific part of a functioning human body.
   d. deoxyribonucleic acid.

2. To say that a characteristic is multifactorial means that:
   a. many genes are involved.
   b. many environmental factors are involved.
   c. many genetic and environmental factors are involved.
   d. the characteristic is polygenic.

3. The best way to differentiate between genetic and environmental influences is to study children who have:
   a. different genes and environments.
   b. the same genes and environments.
   c. similar genes and similar families.
   d. the same environments, but different genes.

4. If an ovum is fertilized by a sperm bearing a Y chromosome:
   a. a female embryo will develop.
   b. a male embryo will develop.
   c. Klinefelter's syndrome will occur.
   d. a miscarriage will occur.

5. When a zygote splits and the two identical halves develop independently, the resulting birth will produce:
   a. dizygotic twins.
   b. monozygotic twins.
   c. fraternal twins.
   d. trizygotic twins.

6. A person with the condition called "mosaicism" has a greater chance of producing:
   a. a baby with a club foot.
   b. degenerating ova or sperm.
   c. a gamete with too many or too few chromosomes.
   d. destructive genes.

7. Shortly after the zygote is formed, it begins a process of duplication and division. Each resulting new cell has:
   a. the same number of chromosomes as the zygote.
   b. half the number of chromosomes as the new zygote.
   c. twice, then four times, then eight times the number of chromosomes as the zygote.
   d. all the chromosomes except those that determine sex.

8. A person who is a "carrier" of a genetic disorder but does not suffer from it manifests the abnormality:
   a. only in his or her genotype.
   b. only in his or her phenotype.
   c. in either the genotype or the phenotype.
   d. in both the genotype and the phenotype.

9. Each human cell contains:
   a. 23 chromosomes.
   b. 23 pairs of chromosomes.
   c. 46 pairs of chromosomes.
   d. 46 genes.

10. Two people with brown eyes have the same _____; however, they may have different _____.
    a. phenotype . . . genotypes.
    b. genotype . . . phenotypes.
    c. recessive genes . . . dominant genes.
    d. dominant genes . . . recessive genes.

11. Genetic testing and counseling are recommended:
    a. when a couple already has a child with a genetic disorder.
    b. when relatives of a couple have genetic problems.
    c. when both parents are from the same genetic stock.
    d. in all of the above situations.

12. The experimental technique in which normal genes are cultivated and exchanged for abnormal genes is called:
    a. chorionic villi sampling.
    b. gene mapping.
    c. amniocentesis.
    d. gene replacement therapy.

13. In the audio program, Don and Karen decided to have a second child, even though Don was a carrier of a genetic condition in which there was a(n):
    a. dominant gene for Down syndrome.
    b. genetic incompatibility to Karen.
    c. excess of amniotic fluid.
    d. chromosome translocation.

14. Concerning gene mapping and gene replacement therapy, which of the following is true?
    a. At the present time, more advances have been made in gene mapping than in gene replacement therapy.
    b. More advances have been made in gene replacement therapy than in gene sampling.
    c. The technology is not yet sophisticated enough for gene replacement to be used.
    d. All of the above are true.

15. The prenatal genetic diagnosis technique that can be done much earlier in the pregnancy and produces faster results than amniocentesis is:
    a. chorionic villus sampling.
    b. gene replacement sampling.
    c. Huntington's technique.
    d. karyotyping.

LESSON 3 EXERCISE: GENETIC LEGACIES

As you heard in the audio program, Don was a carrier of a genetic condition called a **chromosome translocation,** in which a fragment of one chromosome breaks off and becomes attached to another. In Don's case, chromosomes 3 and 15 were involved. Don inherited one normal chromosome 3 (Normal 3), one abnormally short chromosome 3 (Short 3), one normal chromosome 15 (Normal 15), and one abnormally long chromosome 15 (Long 15) that contained the fragment from chromosome 3. Although he was a carrier of the genetic disorder, it did not appear in his phenotype. The disorder is manifest in a person's phenotype only if the abnormally long chromosome is transmitted without the abnormally short one—or vice versa. This condition was manifest in Don's brother and niece, who are severely retarded.

It is important to note that not all genetic disorders are of this kind. Chromosomal translocations of a different type are involved in disorders such as Down syndrome. These disorders are different again from **single gene disorders** such as Huntington's disease, cystic fibrosis, sickle cell anemia, Duchenne's muscular dystrophy, and hemophilia. The varieties and causes of many of these genetic disorders are discussed in Chapter 3 of *The Developing Person Through the Life Span, 2/e.*

If Karen and Don became parents, Karen would transmit normal chromosomes. Their child's condition would therefore depend on Don. There would be four possible legacies. To understand these legacies, fill in the boxes of the figure that follows. The first is filled in for you. Then see if you can answer the questions that follow. Send the completed exercise to your instructor.

LESSON GUIDELINES

Audio Question Guidelines

1. **Genotype** refers to an individual's entire genetic makeup, whether or not these genes are expressed outwardly.

    When the trait is apparent, it means the genes have expressed themselves in the person's **phenotype.**

2. **Amniocentesis** is a prenatal diagnostic test that can reveal genetic problems in a fetus months before birth.

    A needle is inserted into the uterus, where the fetus floats in a sac filled with **amniotic fluid.**

    Amniotic fluid contains cells shed by the fetus. A sample of fluid is withdrawn and the chromosomes of the cells are magnified, photographed, and examined for chromosomal abnormalities.

    In the case of Karen and Don's unborn fetus, an examination of its chromosomes showed that the child would not only be a carrier of the genetic abnormality but would also develop the characteristic physical and mental abnormalities in its phenotype. On the basis of this information they decided to terminate the pregnancy.

3. When any healthy young couple undertakes a pregnancy, there is a two- to three-percent baseline risk that there could be an abnormal outcome, such as a significant birth defect resulting in mental retardation, or a shortened life span.

4. **Prenatal diagnosis:** As in Karen and Don's case, prenatal diagnosis and counseling are available to prospective parents concerned that a genetic condition may run in their family.

    Pediatric genetic counseling is available to families with children born with significant birth defects, and children who, though apparently healthy at birth, later show a decline in development that suggests a genetic condition.

    Counseling for **adult-onset conditions** is available for genetic conditions such as Huntington's disease, presenile dementia, certain neuromuscular disorders, and other problems that do not begin to be expressed until the adult years.

5. **Chorionic villi sampling:** A catheter is inserted into the placenta parallel to the wall of the uterus. Through the catheter is removed a sample of the villi—fingerlike projections that dip into the lining of the uterus.

    The villi are composed of fetal tissue, the cells of which are dividing so rapidly that their chromosomes can be examined directly.

    Two advantages of this technique over amniocentesis are that it can be done earlier in the pregnancy, and that the results are available to parents sooner.

    Gene mapping: Gene mapping refers to techniques that identify the abnormal genes specifically causing a genetic disorder.

Through gene mapping, abnormal genes have been identified for Huntington's disease, cystic fibrosis, sickle cell anemia, and other genetic conditions.

Gene replacement therapy: Gene replacement therapy refers to the experimental process in which "good" genes are cultivated and substituted for abnormal genes in a diseased person's tissue.

Still in the experimental stages, and not without possible negative effects, gene replacement therapy may eventually be available to treat some disorders.

Textbook Question Guidelines

1. About once in every 270 pregnancies, a **zygote** splits into two identical cells that develop into identical, or **monozygotic,** twins who are the same sex, look alike, and share all other inherited characteristics.

   Dizygotic, or fraternal, twins develop from two separate zygotes and share no more genes than any other siblings.

   Research using twins is useful for determining the relative contributions of genes and the environment: for any given characteristic, if monozygotic twins are much more alike than dizygotic twins, genes rather than environment would appear to determine that characteristic.

2. Most human characteristics are **multifactorial,** that is, they are the product of the interaction of many genetic and environmental factors. Even physical characteristics, such as height, are not completely determined by genes. Psychological characteristics, such as intelligence, personality, and mental illness, are even more clearly multifactorial.

   Because changes in the environment may diminish or exacerbate the effects of inherited predispositions, one must be cautious in assuming that a genetic predisposition implies an inevitable expression.

   The relationship between genes and environment also varies from person to person.

3. Between 100,000 and 150,000 children are born in the United States each year with a chromosomal abnormality. Many involve the twenty-third pair—the sex chromosomes:

   Kleinfelter's syndrome: XXY chromosomal pattern results in a boy with retardation in language skills and undeveloped secondary sex characteristics.

   XYY chromosomal pattern: unusually aggressive boy with mild retardation.

   XXX chromosomal pattern: girl retarded in most intellectual skills.

   XO (only one sex chromosome): short, mildly retarded girl who fails to develop secondary sex characteristics.

   Other abnormalities occur when a sperm or ovum contains 23 autosomes instead of 22.

   Down syndrome children usually have distinctive features of the eyes, nose, and tongue. They often suffer heart defects and have slower physical and intellectual development.

4. The causes of genetic problems include:

Mosaicism, in which one parent has an extra or a missing chromosome in some cells.

Middle-aged parents, who have more offspring with chromosomal problems.

The **fragile X syndrome,** in which the X chromosome does not transmit genetic instructions effectively.

Harmful genes that are present in every individual's genotype and that seriously affect the phenotype of about one infant in thirty.

Answers to Testing Yourself

1. **b.** A zygote is a fertilized egg; DNA *is* deoxyribonucleic acid. Genes contain the instructions that differentiate cells into specific parts of the human body. (textbook, page 55)

2. **c.** Most important human characteristics are the product of the interaction of many genetic *and* environmental factors. (textbook, pages 62–63)

3. **d.** Studies of twins are useful because they permit researchers to study the effects of heredity and environment, while holding one of the two factors constant. (textbook, pages 58–59)

4. **b.** In males, the twenty-third pair of chromosomes is *XY;* in females the pattern is *XX.* (textbook, page 56)

5. **b.** A zygote is a fertilized egg. Identical twins develop from a single zygote; fraternal twins develop from two separate zygotes created by the fertilization of two separate ova. (textbook, page 57)

6. **c.** Mosaicism is the condition in which a person has an extra or a missing chromosome in some cells. Such a person has a high probability of contributing an abnormal reproductive cell to the formation of a zygote. (textbook, page 65)

7. **a.** Through the process of cellular division called mitosis each new cell has the same genetic information that was contained in the zygote. (textbook, pages 55–56)

8. **a.** A carrier manifests the abnormality only in his or her genotype. (audio program; textbook, page 60)

9. **b.** The child receives 23 pairs of chromosomes, one of each pair from the father and one from the mother. (audio program; textbook, page 56)

10. **a.** Genotype refers to an individual's genetic makeup; phenotype refers to the person's observable characteristics. (audio program; textbook, page 60)

11. **d.** In each of these situations there is a strong possibility that a defective genetic condition exists. (audio program; textbook, pages 70–71)

12. **d.** Although still in the experimental stage, gene replacement therapy may someday be a viable form of treatment for many genetic disorders. (audio program)

13. **d.** A fragment from chromosome 3 had broken off and become attached to Don's chromosome 15. (audio program)

14. **a.** More advances have been made in genetic mapping. (audio program)

15. **a.** Chorionic villi sampling can be done as early as 9 weeks into a pregnancy. (audio program; textbook, page 71)

References

Plomin, R., DeFries, J. C., & McClearn, G. E. (1980). *Behavioral genetics: A primer*. San Francisco: Freeman.

A helpful, readable introduction to genetic research.

Prenatal Development and Birth

AUDIO PROGRAM: When to Have a Baby

ORIENTATION

Is 21 years of age too young to become a parent? Is 36 years of age too old? There are no simple answers to these questions. As Lesson 4 explains, it all depends on the settings of three developmental clocks that tick through the seasons of life, from the very beginning to the very end. The first is the **biological clock,** which is the body's timetable for growth and decline. The second is the **social clock,** a culturally set timetable that establishes when various events in life are most appropriate. The third is the **psychological clock,** our personal timetable of readiness for life's milestones.

One of the themes of *Seasons of Life* is that the diversity of life-span development is due in part to the fact that the social and psychological clocks are not set the same for everyone. Just as each culture, subculture, and historical period establishes its own social clock, so each individual establishes his or her own psychological clock on the basis of individual life experiences.

In audio program 4, "When to Have a Baby," two couples about to give birth to their first child discuss their impending parenthood. Because one of the expectant mothers is 21 and the other 36, the life-span consequences of their "early" and "late" births will be very different. Their stories, illuminated by the expert commentary of sociologist Alice Rossi and anthropologist Jane Lancaster, illustrate how the three developmental clocks influence the timing of births. Although the clocks come into play in every major transition of life, they are not always in synchrony, and their settings have been changed over the course of history.

The birth of a child is one of life's most enriching experiences. Nine months of prenatal development culminate in the expectant couple assuming a new and demanding role as parents, and being transformed from a couple to a family. But, as discussed in Chapter 4 of the textbook, parental responsibilities begin long before birth, during the period of prenatal development. This development is outlined together with a description of some of the problems that can occur in the perinatal period, including preterm birth, low birth weight, and prenatal exposure to disease, drugs, and environmental hazards.

As the program opens we hear the voices of the two couples pondering their imminent transition to parenthood.

LESSON GOALS

By the end of this lesson you should be prepared to:

1. Differentiate the biological clock, the social clock, and the psychological clock, and discuss their significance in the timing of births.

2. Discuss how the timing of births and the setting of the three developmental clocks have changed over the course of human history.

3. Outline the rapid and orderly development that occurs between conception and birth.

4. Explain the general risk factors and specific hazards that may affect prenatal development.

5. Discuss the psychological impact of pregnancy and birth on expectant parents.

Audio Assignment

Listen to the audio tape that accompanies Lesson 4: "When to Have a Baby." Write answers to the following questions. You may replay portions of the programs if you need to refresh your memory. Answer guidelines may be found in the Lesson Guidelines section at the end of this chapter.

1. Differentiate the biological clock, the social clock, and the psychological clock and discuss their significance in development through the life span.

2. Discuss whether the settings of the three developmental clocks are different for different generations.

3. Explain how the pattern and timing of childbearing changed as humans shifted from a hunting-and-gathering society to a modern society.

4. Discuss some of the life-span consequences of births that occur early and late in parents' lives.

Textbook Assignment

> Read Chapter 4: "Prenatal Development and Birth," pages 77–99 in *The Developing Person Through the Life Span, 2/e.*
>
> Write your answers to the following questions. Refer back to the textbook, if necessary. Answer guidelines may be found in the Lesson Guidelines section at the end of this chapter.

1. Describe the significant developments that occur in each of the three periods of prenatal development: the germinal period, the period of the embryo, and the period of the fetus.

   Germinal period

   Period of the embryo

   Period of the fetus

2. Identify several types of teratogens and describe their possible effects on the developing embryo or fetus.

3. Identify the major problem for infants who undergo stressful birth. Explain how obstetricians know when the fetus is experiencing stress, and describe the most likely medical interventions in such a case.

4. Discuss the causes and complications of low birth weight and preterm birth.

5. Identify several psychological factors that influence parents' overall experience of birth and describe several approaches to improving this experience.

6. Discuss whether there is a critical period in development for the formation of the parent–infant bond.

Testing Yourself

After you have completed the audio and text review questions, see how well you do on the following quiz. Correct answers, with text and audio references, may be found at the end of this chapter.

1. The social clock tells us:
   a. the age at which having a child becomes biologically feasible.
   b. the average age for having a first child.
   c. the appropriate or "best" age for having a child in our society.
   d. the age at which having a child best correlates with the parents' well-being later on.

2. In contrast to hunter–gatherer women, sedentary women tend to have:
   a. more children.
   b. fewer children.
   c. fewer menstrual cycles.
   d. a shorter fertile period.

3. The audio program states that for the typical American woman today the biological and social clocks are out of sync. Which of the following statements explains why this is so?
   a. The average teenager today is sexually mature before she is psychologically interested in sexual activity.
   b. Although menarche occurs at a younger age than ever before, it takes longer than ever to achieve the social status of an adult.
   c. Most women today assume the social role of adults before their reproductive systems are optimally suited for childbearing.
   d. Because of the widespread use of oral contraceptives, the biological clock that governs menstruation has effectively been halted.

4. Two patterns of childbearing are common today. The one associated with "early" births favors the _____ clock, while the one associated with "late" births favors the _____ clock.
   a. social . . . biological
   b. social . . . psychological
   c. biological . . . social
   d. psychological . . . social

5. "Sedentism" refers to:
   a. the tendency for menarche to occur at an earlier age in recent years.
   b. the process by which the biological clock governs the optimal years for childbearing.
   c. the tendency of less active women to have a later menarche.
   d. the shift in human social organization from a nomadic life to a village-dwelling society.

6. The third through the seventh week after conception is called the period of the:
   a. embryo.
   b. ovum.
   c. fetus.
   d. germinal.

7. The most critical factor in attaining viability (between the 20th and 26th week) is the development in the fetus of the:
   a. placenta.
   b. eyes.
   c. brain.
   d. skeleton.

8. The neonate's heart rate, breathing, muscle tone, circulation, and reflexes are usually measured on a ten-point test known as the:
   a. fetal monitor.
   b. critical period.
   c. Apgar scale.
   d. Lamaze scale.

9. A characteristic of a teratogen is that it:
   a. cannot cross the placenta during the first trimester.
   b. is usually inherited from the mother.
   c. can be counteracted by good nutrition most of the time.
   d. may be a virus, drug, or radiation.

10. Because most body organs form during the first two months of pregnancy, this time is sometimes called the:
    a. period of teratology.
    b. genetic danger term.
    c. proximo-distal period.
    d. critical period.

11. Studies of fathers present during delivery found that:
    a. most became emotionally disoriented.
    b. most were glad they had been there.
    c. mother–infant bonding was disrupted.
    d. most would not repeat the experience.

12. The idea of a parent–infant bond in humans, that is, a bond that might have to be formed a short time after birth, arose from:
    a. observations in the delivery room.
    b. data on adopted infants.
    c. studies in animals.
    d. studies of disturbed mother–infant pairs.

13. During the hours after birth, babies born to mothers who received no medication tend to be:
   a. larger and heavier than average.
   b. more alert and responsive to their surroundings.
   c. less alert and responsive to their surroundings.
   d. more likely to experience postnatal depression.

14. An example of proximo-distal development would be the:
   a. development first of the most vital internal organs and body parts.
   b. development first of the upper arms, then lower arms, then hands, then fingers.
   c. development of legs prior to the development of arms.
   d. presence of a primitive tail structure.

15. An important occurrence of the last trimester is the:
   a. beginning of the infant's heartbeat.
   b. formation of the sex organs.
   c. beginning of the change of cartilage into bone.
   d. rapid development of the brain.

LESSON 4 EXERCISE: SAYING WHEN

The three developmental clocks come into play in every major transition of life. No transition is greater than the change from being pregnant to being a parent. The setting of the developmental clocks can affect the timing of births, the adjustment of first-time parents to their new roles, and can have long-term life-span consequences on both parents and their children.

The stories of the two couples introduced in audio program 4 illustrate some of these effects and contrast two new patterns of childbearing in our species. Shelley and her husband Charles gave birth when Shelley was 21. Shelley's pregnancy was "on time" biologically, but "off time" socially and psychologically. The pregnancy was unexpected and came at a time when the young couple was still establishing their own relationship, completing their educations, and struggling to make a living.

Brett and Henry's child was born when Brett was 36. This biologically "late" birth is an example of an increasingly common pattern of childbearing that favors the social clock by allowing parents to establish careers and improve their financial security before having children. Brett and Henry's birth may have been "off time" biologically, but it was "on time" psychologically and "on time," or perhaps even a little late, in terms of the social clock.

The experts in the audio program point out that unlike the biological clock, which changes little from generation to generation, the social clock and psychological clock can be reset. The settings of these clocks reflect each individual's culture, historical context, and life experiences.

To help you integrate the material in Lesson 4 into an actual life story, write answers to the questions on the **Exercise Response Sheet** on the following page. Return the completed sheet to your instructor. Before completing the exercise, you may find it helpful to review Lesson Guidelines 1–4 to make sure you understand the differences and interrelationships of the three developmental clocks.

LESSON GUIDELINES

Audio Question Guidelines

1. The **biological clock** governs physical development through the various mechanisms of heredity and physiology that program growth, fertility, and aging. For example, for women, the biologically optimal period for having a child is between 22 and 32 years; for men the range is between 22 and 40 years.

   The **social clock** is a culturally set timetable that establishes when various events and behaviors in life are appropriate and called for. One of the reasons that development is so diverse is that the social clock is not the same for everyone. Each culture, subculture, and generation has a somewhat different social clock.

   The **psychological clock** represents each person's inner timetable of development. It is the individual's way of determining when he or she is ready to marry, to become a parent, and to make the other transitions inherent in development.

2. Unlike the two other developmental clocks, which can be reset based on the individual's life experiences, culture, and generation, the **biological clock** is relatively immutable. For most life events, however, the biological clock specifies a normal *range* of time rather than a precise moment. For example, although the biological clock was unchanged in our transition from a hunter–gatherer society to a sedentary society, because of better nutrition and living conditions in modern times, the average age of menarche (12.5 years) has dropped to the lower end of the range set by the biological clock.

   At the same time that young people are becoming sexually mature at younger ages, the social clock has moved in the opposite direction: it takes longer and longer to achieve the status of an adult in our society. The result is that today's society has created a lengthy period in which an individual may have the reproductive capacity of an adult but the social role of a child.

   Since it is set according to each individual's life experiences, the psychological clock shows the greatest diversity of the three developmental clocks. Because she lives at a time when the biological and social clocks are out of sync, the typical American woman today finds it very difficult to decide when to have a child. She will find that the burden falls on her individual psychological clock.

3. As recently as 10,000 years ago, our ancestors lived as hunter–gatherers. Unlike modern women, hunter–gatherer women nursed their children for three or four years. This continuous nursing tended to suppress ovulation and limit the number of children born. With the development of agriculture and the shift from a nomadic to a **sedentary** lifestyle, the numbers of children born increased. Another change is that the average age of menarche—the beginning of the menstrual cycle—dropped from about age 16 in hunter–gatherer women to 12.5 years in today's sedentary women.

   For hunter–gatherer women, the biological and social clocks were in sync: by the time they had their first child they had already assumed their social role as adults. Because the social and biological clocks are not in sync today, two new patterns of childbearing have emerged. One, favoring the biological clock, is the bearing of children very early in the life cycle (before the age of 21). The second, favoring the social clock, is exemplified by women who postpone having children until age 35 or later in order to become established in a career.

4. During most of human history the peak reproductive years were between 20 and 30. The two new patterns of childbearing that have emerged today— "early" births and "late" births—are atypical of our species' history.

As in the case of Shelley and Charles whom we met in the audio program, an "early" birth is one that may be on time biologically but off time socially and psychologically. When the three developmental clocks are out of sync, a difficult period of adjustment may follow.

Once one moves outside of the optimal biological range for having a baby— between 22 and 32 for women and between 22 and 40 for men—there is a greater risk of physiological impairment in the infant. Many early births (to mothers under age 16 or 17) result in low birth weight babies. Late births (to mothers over the age of 38) are associated with increased risk of physiological impairment resulting from conditions such as Down syndrome.

One of the major reasons for the postponement of birth has been among better-educated segments of the population, who wish to complete their training or become established in a career before becoming parents. The late-timed baby may benefit from the parents' greater economic security and maturity.

Because of the ages of the older parents, late-timed births are often only births: the one-child family is becoming more and more common.

Textbook Question Guidelines

1. During the germinal period the one-celled **zygote** multiplies, becoming a mass of cells. Approximately 10 days after conception, **implantation** occurs. The many-celled organism burrows into the lining of the uterus to obtain nourishment and initiate the hormonal changes that will prevent it from being expelled during the next menstrual cycle.

With the formation of the neural tube (the beginning of the brain and spinal cord) at 21 days, the organism is called an **embryo.** The period of the embryo extends from two weeks to eight weeks after conception. Growth is rapid and orderly, with the most vital organs and body parts developing first. The head develops more rapidly than the lower body **(cephalo-caudal development),** and parts closer to the center of the embryo develop before those farther away **(proximo-distal development).** Also important is the rapid growth of the placenta—the organ that provides a mechanism of transfer and filtration of oxygen and nourishment between the mother and embryo.

The period of the **fetus** (two months to birth) includes the second and third **trimesters** (three-month periods) of the pregnancy. During the third month the muscles develop, major organs complete their formation, the sex organs begin to take shape, and the fetus can move almost every part of its body. During the second trimester hair, eyebrows, fingernails, and teeth buds form and grow. The fetal heartbeat can be heard, brain development is extensive, and, at 20–26 weeks, the **age of viability** is reached. During the third trimester the lungs and heart become capable of sustaining life without the placenta, patterns of sleeping and waking emerge, and the fetus gains approximately 5½ pounds.

2. **Teratogens** are substances that can cross the placenta and harm the developing baby. A particular organ or body part is most susceptible to damage during its **critical period,** which is most likely to occur during the first trimester, when most of the major organ systems are being formed.

   Teratogens include **diseases** such as rubella, toxoplasmosis, syphilis, mumps, and genital herpes. If the mother contracts rubella early in pregnancy, for example, many birth handicaps—including blindness, deafness, heart abnormalities, and brain damage—may result.

   Many **drugs,** including streptomycin, Valium, most hormones, and Thorazine, have proven teratogenic effects. Even nonprescription drugs, such as aspirin and antacids, have been implicated in birth defects. About a third of babies born to alcoholic mothers suffer from **fetal alcohol syndrome;** resulting deformities may include a small head, malproportioned face, and mental retardation.

   Environmental hazards for prenatal development include overexposure to radiation (x-rays) and pollutants, including carbon monoxide, lead, mercury, and PCBs.

3. A long, complicated, and stressful birth may result in lifelong handicaps for the newborn. The major problem in such births is **anoxia,** or lack of oxygen. Prolonged anoxia can cause brain damage and even death, especially if the fetus is **preterm.**

   Monitoring of the fetus's heartbeat and the mother's contractions allows obstetricians to determine whether labor is too stressful for the mother or the fetus. The birth process may be hastened by the performance of a **Cesarean section,** a surgical incision made in the mother's abdomen through which the baby is removed.

   In recent years a growing number of doctors and expectant parents have reacted against the depersonalization and medicalization of hospital births.

4. **Full term** newborns have an average weight of 7½ pounds.

   Approximately 7 percent of newborns in the United States weigh less than 5½ pounds and are designated **low-birth-weight** infants. Most are also **preterm,** being born more than 3 weeks before they are due.

   Preterm birth is the leading cause of newborn death. Other problems faced by the preterm newborn include immature reflexes, increased risk of cerebral palsy, blindness, and the need for extra care and soothing environmental stimulation.

   Factors associated with low birth weight include malnourishment of the mother, improper function of the placenta, exposure to teratogens, prenatal infections, and genetic handicaps.

   Lower-class and teenage mothers are more likely to give birth to preterm and/or low-birth-weight babies.

5. Psychological factors that influence the couple's birth experience include the relationship between the mother and father, the couple's willingness to assume the responsibilities of parenthood, and the social support available to both mothers- and fathers-to-be.

   Childbirth classes, such as those teaching the **Lamaze method,** are helpful in reducing the fear and tension that many parents feel prior to the birth.

   When both partners attend childbirth classes and are active participants in the birth, the result is less pain, less need for anesthesia, shorter labor, and more positive feelings about the birth, the baby, and themselves.

6. Although research findings are inconclusive, for most experienced mothers with healthy babies, immediate mother–infant contact is not critical for establishing the mother–child bond.

   For first-time mothers or those who are young, poor, or under stress, extended early contact may promote greater maternal attention and affection.

   For fathers, the data are less conclusive: whether a father has extended contact with his newborn does not predict his involvement with the baby in the months following the birth.

Answers to Testing Yourself

1. **c.** The social clock is a culturally set timetable that establishes when various events and behaviors in life are appropriate and called for. (audio program)

2. **a.** The hunter–gatherer society is associated with greater restraint in the production of children than sedentism. This is due, in part, to the tendency of women in hunter–gatherer societies to nurse continuously, which tends to suppress ovulation and prevent pregnancy. (audio program)

3. **b.** Today's society has created a 10-year, or longer, period in which an individual may have the reproductive capacity of an adult but the social role of a child. (audio program)

4. **c.** Early births favor the biological clock by occurring during the optimal period of biological fertility. Late births favor the social clock because they allow the couple to establish careers and attain greater financial security before having children. (audio program)

5. **d.** The shift from a hunter–gatherer to village-dwelling society is referred to as sedentism. (audio program)

6. **a.** The stages of prenatal development are, in order, the germinal period (first two weeks), the period of the embryo (third week through the seventh week), and the period of the fetus (from the eighth week until birth). (textbook, p. 77)

7. **c.** The development of the brain is essential to the regulation of basic body functions. (textbook, p. 80)

8. **c.** The Apgar scale is used to assess quickly the newborn's physical condition. (textbook, pp. 81–82)

9. **d.** Teratogens are substances that cause birth defects, can cross the placenta, and harm the embryo. (textbook, pp. 82–88)

10. **d.** The critical period for each body part is the time when that part is most susceptible to damage. (textbook, p. 83)

11. **b.** While their actual birth experiences vary, many more patients find it positive than negative. (textbook, pp. 95–97)

12. **c.** The formation of a specific bond between mother and newborn occurs in virtually every species of animal. (textbook, p. 97)

13. **b.** Even low doses of medication administered to the laboring mother can make the newborn less alert. (textbook, p. 93)

14. **a.** Proximo-distal development means from "near to far." (textbook, p. 78)

15. **d.** Brain development is particularly rapid during the third trimester. (textbook, p. 80)

References

Kitzinger, Sheila (1985). *Birth over thirty*. New York: Penguin, 1985.

Kitzinger examines various aspects of later pregnancies, including physical risks such as Down syndrome and Cesarean births, as well as psychological aspects such as how to adjust to the role of motherhood during middle age.

The First Two Years: Physical Development

ORIENTATION

This lesson is the first of a three-lesson unit that describes the developing person from birth to age 2 in terms of physical, cognitive, and psychosocial development. Lesson 5 examines physical development.

Physical development during the first two years is so rapid that infants often seem to change before their parents' very eyes. In chapter 5 of *The Developing Person Through the Life Span,* 2/e, Kathleen Berger describes the typical patterns of growth in the body and nervous system and the timetables of **sensory, perceptual,** and **motor-skill** development. Although the developmental sequence is usually the same for all healthy infants, variation in the ages at which certain skills are mastered does occur, in part because development is dependent on the interaction of biological and environmental forces.

Audio program 5, "The Biography of the Brain," continues the stories of the two couples introduced in program 4. Both couples have now had their babies and in this lesson we follow the early months of the babies' physical development. Compared to other mammals, humans are physically quite immature at birth. Evolutionary biologist Stephen Jay Gould and anthropologist Barry Bogin suggest that during the course of human evolution, an increase in the size of the brain and skull required a corresponding reduction in the length of pregnancy. In this way, the infant can be born before its head has grown too large to pass through the birth canal. Brain development influences growth in other ways as well. Attainment of adult body size and sexual maturity are delayed until the brain, too, is almost fully mature. Brain development continues throughout life, increasingly influenced by environmental factors and learning.

Another issue explored in this lesson is the importance of nutrition to the developing brain. In the program, anthropologist Jane Lancaster notes that mothers and babies store body fat in order to meet the nutritional needs of the developing brain. The mother's fat ensures a rich of supply of milk, an ideal food with a special profile of nutrients that exactly matches the developmental needs of the infant.

Neuropsychologist Jill Becker describes brain development at the microscopic level as a process in which individual **neurons** grow and form **synapses** with other neurons. Laboratory research with animals indicates that being raised in a stimulating environment promotes the development of more of these neural connections.

As the program opens we hear one of the couples describe the birth of their first child and the tight fit of their baby's head through the birth canal.

LESSON GOALS

By the end of this lesson you should be prepared to:

1. Describe the normal patterns of physical, brain, and motor-skill development during infancy.

2. Discuss how biological and environmental forces interact in the infant's sensory development, brain maturation, and the acquisition of motor skills.

3. Identify the competing evolutionary pressures that have led some anthropologists to argue that human babies are born "too soon."

4. Outline the nutritional needs of infants during the first year of life. Describe the significance of breast milk and body fat in ensuring adequate nutrition.

Audio Assignment

Listen to the audio tape that accompanies Lesson 5: "The Biography of the Brain."

Write answers to the following questions. You may replay portions of the program if you need to refresh your memory. Answer guidelines may be found in the Lesson Guidelines section at the end of this chapter.

1. Explain why some anthropologists believe that, compared with other mammals, human babies are very immature at birth.

2. Discuss the significance of breast milk and body fat in meeting the nutritional needs of the newborn.

3. Describe the ways in which the nervous system matures during childhood.

4. Explain how the needs of the brain affect development in other systems of the body.

Textbook Assignment

> Read Chapter 5: "The First Two Years: Physical Development," pages 103–119 in *The Developing Person Through the Life Span, 2/e.*
> Write your answers to the following questions. Refer back to the textbook, if necessary. Answer guidelines may be found in the Lesson Guidelines section at the end of this chapter.

1. Describe the size and proportions of the average newborn's body, how these change during infancy, and how they compare to those of an average adult.

2. Discuss sudden infant death syndrome and the major risk factors associated with it.

3. Describe the sensory and perceptual abilities of the human infant.

4. Outline the basic pattern of motor-skill development in humans.

5. Compare the advantages and disadvantages of breast-feeding and bottle-feeding.

6. Discuss the causes and results of malnutrition among children in developing countries and the United States.

Testing Yourself

> After you have completed the audio and text review questions, see how well you do on the following quiz. Correct answers, with text and audio references, may be found at the end of this chapter.

1. According to experts in the audio program, why are human babies born so physically immature?
   a. As a result of better nutrition, the biological clock that times gestation has been reset.
   b. If prenatal development continued longer, infants' large heads would not fit through the birth canal.
   c. Humans are much less active than other mammals.
   d. All of the above are reasons why babies are born so physically immature.

2. What is the correct order in which physical maturation occurs in the human body?
   a. Brain, body, reproductive system
   b. Body, brain, reproductive system
   c. Body, reproductive system, brain
   d. Brain, reproductive system, body

3. Approximately 90 percent of the brain's growth is completed by the age of:
   a. 1 month.
   b. 6 months.
   c. 1 year.
   d. 5 years.

4. Laboratory research with animals has shown that one effect of being raised in a stimulating environment is:
   a. an increase in the number of neurons in the brain.
   b. a shortening of the time required for the brain to reach its full capacity.
   c. a reduction in the number of "extra" nerve cells that die.
   d. a reduction in nerve cell insulation, which tends to slow neural communication.

5. Compared to that of other species, human milk has a high content of _____, which especially promotes development of _____.
   a. protein . . . the brain.
   b. protein . . . muscle.
   c. sugar . . . muscle.
   d. sugar . . . the brain.

6. The cephalo-caudal pattern of physical development explains why:
   a. the newborn's feet are longer than the hands.
   b. babies are born with a high percentage of body fat.
   c. the newborn's head is more fully developed than the rest of the body.
   d. the internal organs develop earlier than the limbs.

7. Animals that are prevented from using their senses normally experience:
   a. a temporary impairment in sensory abilities.
   b. a temporary impairment in perceptual abilities.
   c. permanent sensory and perceptual impairment.
   d. no significant sensory or perceptual impairment.

8. Which of the following is the best example of sensation?
   a. An infant prefers looking at her mother's face.
   b. A researcher strikes a tuning fork to test a newborn's hearing.
   c. A child identifies the name of a popular song on the radio.
   d. Nerve cells in an infant's visual system react to a change in light intensity.

9. Infants are correctly referred to as "toddlers" when they begin to:
   a. talk.
   b. stand by themselves.
   c. reach for and grasp objects.
   d. crawl.

10. Reflexes are:
    a. motor skills that are mastered through practice.
    b. involuntary motor responses to environmental stimuli.
    c. fully developed at birth, even in preterm infants.
    d. motor skills that disappear once the nervous system is mature.

11. The protein deficiency that results in swelling of a child's face and body is known as:
    a. marasmus.
    b. veganism.
    c. kwashiorkor.
    d. shingles.

12. Most problems in the nutrition of American children are caused by:
    a. insufficient calorie intake.
    b. unhealthy dietary practices.
    c. a diet too high in protein.
    d. poor food preparation practices.

13. Compared with bottle-fed infants, breast-fed infants:
    a. gain weight faster.
    b. gain weight more slowly.
    c. become socially secure earlier.
    d. have fewer allergies and digestive problems.

14. Of the following, which sense is the *least* developed at birth?
    a. Smell
    b. Taste
    c. Hearing
    d. Vision

15. Which of the following is *not* identified in the textbook as a risk factor for the sudden infant death syndrome?
    a. Being a first-born child
    b. Being born to a mother younger than 20
    c. Being male
    d. Weighing under 5.5 pounds at birth

LESSON 5 EXERCISE: GROWTH RATES AND MOTOR-SKILL DEVELOPMENT IN THE FIRST TWO YEARS

The text notes that weight gain and growth in early infancy are astoundingly rapid. The proportions of the human body change dramatically with maturation, especially in the first year of life. The growth that changes the baby's body shape follows the head-downward (cephalo-caudal) and center-outward (proximo-distal) direction of development. For instance, the percentage of total body length below the belly button is 25 percent at two months past conception, about 45 percent at birth, 50 percent by age 2, and 60 percent by adulthood. Questions 1–5 of this exercise will help you to gain some understanding of how rapid this growth is by projecting the growth patterns of the infant onto an adult, such as yourself or a friend.

One way to understand how infants master basic motor skills is to try to relearn them as an adult. A good place to start is with the mastery of motor skills involved in picking up objects. Questions 6–10 of this exercise are designed to help you "feel" the progressively finer coordination achieved by the infant by working through the stages of development. Return your completed **Exercise Response Sheet** to your instructor.

LESSON GUIDELINES

Audio Question Guidelines

1. By mammalian standards, the nine-month gestation period in humans is relatively short. As a result, the human infant is exceedingly helpless and relatively less developed compared to most other animals. The relatively short gestation period may be a response to the evolution of a large brain, which, while providing humans with a high level of intelligence and cognitive flexibility, results in a large skull that makes birth difficult.

   Because of bipedalism (the upright posture of humans), the duration of prenatal development is somewhat of a compromise. Babies are born at a point when the brains are as big as they can be without requiring a wider birth canal, which would impair a female's ability to walk.

2. Compared to other primates, human infants are remarkably fat when they are born. Fat ensures that the developing brain—which has attained only one-quarter of its mature size at birth—is adequately nourished.

   Even before infants are born, their mothers store fat for them. The mother's fat ensures a rich supply of breast milk for nursing. Each species' milk has a special profile of nutrients that exactly matches the species' specific developmental needs. Human milk is high in sugar, which provides energy for the baby's rapidly developing brain.

3. The brain consists of billions of nerve cells called neurons. All the neurons a human will ever possess are already present three months after conception. Some of the neurons will grow while others will die. Neuropsychologists estimate that humans start out with 20 to 40 percent more neurons than they end up with.

   During the first two years of life, neurons in the human brain become coated with insulation (which speeds their chemical communication) and hook up with other neurons by forming connections called **synapses.** The excess of neurons and synaptic connections creates redundancy that gives humans greater flexibility and insurance: if some neurons die, others can take over their functions.

   While 90 percent of the brain's growth is completed by age 5, synapses will be made and broken throughout life. Laboratory animals respond to stimulating environments by developing more synapses than those raised in normal or deprived environments. Even in old age, nerve cells respond to stimulation by growing extra connections.

4. The biological clock governs every aspect of physical growth. One of the peculiarities of human growth is that we grow slowly. Another is that we do not grow evenly. Development of the body and reproductive system is paced by development of the brain.

Brain development is rapid, with 90 percent of growth completed by age 5.

Body growth from birth to two years is very rapid and then levels off to a rate of 3 to 4 inches a year until puberty, when growth in height and weight again accelerates.

The curve of growth for the reproductive system is very flat until the age of 12 or 14.

Textbook Question Guidelines

1. The average North American newborn measures 20 inches and weighs about 7 pounds. The newborn seems top-heavy in body proportions, with its head being one-fourth of total body length (in comparison to one-eighth of body length in an adult).

   Growth is "down (**cephalo-caudal**) and out (**proximo-distal**)."

   By 1 year of age body weight has tripled to an average of 22 pounds. Body length has increased to almost 30 inches.

   Growth slows during the second year. By age 2, the toddler's body weight averages 30 pounds (one-fifth adult weight) and body length ranges from 32 to 36 inches, or one-half adult height.

2. One in every 350 infants dies from sudden infant death syndrome (SIDS). The actual cause of SIDS is unknown.

   SIDS is more common in mothers who have already lost a child to SIDS, are younger than 20, have O, B, or AB blood type, smoke, have low incomes, have anemia, have a pregnancy that lasts less than 8 months, and have urinary or respiratory infections.

   SIDS occurs more frequently in infants who are male, later-born, weigh less than 5.5 pounds at birth, had an Apgar score of 7 or lower 1 minute after birth, were born during the winter, are between 2 and 4 months of age, and have a cold.

3. At birth, both **sensation** and **perception** are apparent. Newborns can see, hear, smell, taste, and respond to pressure, motion, temperature, and pain.

   Vision is the least well developed of the senses. Newborns can focus well on objects that are about 10 inches away (20/600 vision).

   By 6 months visual acuity approaches 20/20.

   By 3 months of age infants respond to their mother's facial expressions and recognize her photograph.

   Hearing is comparatively acute in the newborn. By 1 month, infants can differentiate their mother's voice from the voices of other women and can perceive differences between very similar sounds.

4. The newborn's motor ability is limited to **reflexes**, including those that maintain adequate oxygen, such as the **breathing reflex;** those that help maintain body temperature, such as crying and shivering; those that ensure adequate nourishment, such as the **sucking reflex, rooting reflex,** swallowing, and crying. Other reflexes may not be necessary for survival, but are useful

as signs of normal brain and body function. The **Brazelton Neonatal Behavioral Assessment Scale** rates 26 reflexes that are a gauge of the newborn's physical condition.

The sequence of **motor-skill** development is universal and follows the same cephalo-caudal and proximo-distal patterns of physical growth.

By 6 months most babies can reach, grab, and hold onto dangling objects. Between 9 and 14 months babies become able to use the thumb and forefinger to grasp objects (pincer grasp).

The average child can stand with support at 5 months, walk with assistance at 9 months, and walk well unassisted at 12 months.

5. Breast milk is the ideal food for most babies. It is always sterile, at body temperature, contains more essential vitamins and iron than cow's milk, is more digestible, and provides the infant with the mother's immunity to disease.

Despite the advantages of breast-feeding, most American babies are bottle-fed.

For reasons based on convenience, cultural ˜˙˙ ˌ, and protection of the infant from drugs and other potent˙ ˙ ˌottle-feeding may sometimes be the wisest choice.

Neither method of feeding ne ˌarily guarantees, or inhibits, good mother–child relationships.

6. Good nutrition after weaning requires a variety of foods. During the first two years infants need about 50 calories per day per pound of body weight (110 per kilogram).

The primary cause of **malnutrition** in developing countries is early cessation of breast-feeding and the feeding of infants with contaminated or improperly prepared formulas.

Moderate malnutrition may disrupt normal patterns of behavior, such as social responsiveness. When not fatal, long-lasting malnutrition can cause intellectual as well as physical deficits.

Protein-calorie deficiency can cause **marasmus** in infants and **kwashiorkor** in toddlers.

Problems in the nutrition of American toddlers are usually due to poor food practices rather than inadequate supply of nutrients.

Answers to Testing Yourself

1. **b.** Human babies are born at a time when their brains are as big as they can be without requiring a wider birth canal, which would limit our ability to walk. (audio program)

2. **a.** Our slow, uneven growth—first the brain, then the body, then the reproductive system—is directed by the pre-eminent role of the brain. (audio program)

3. **d.** Three-quarters of brain development occurs after birth; this may be the reason why humans store fat: to ensure that the brain has adequate nutrition. (audio program)

4. **c.** Environmental stimulation throughout the life span helps individuals, and their brains, grow. (audio program)

5. **d.** Human milk has a high sugar content, which provides energy for the baby's rapidly developing brain. (audio program)

6. **c.** Cephalo-caudal means "from the top down." (textbook, p. 104)

7. **c.** Animals that are prevented from using their senses or moving their bodies in infancy experience permanent sensory and perceptual impairment. (textbook, p. 106)

8. **d.** Sensation occurs when a sensory system, such as the eye, responds to a particular stimulus, such as light. (textbook, p. 108)

9. **a.** Children are correctly referred to as **infants** until they begin to talk. (textbook, p. 114)

10. **b.** Because reflexes are involuntary physical responses, tests of reflexes are often used to assess the newborn's physical condition. (textbook, p. 111)

11. **c.** Although children with kwashiorkor are severely malnourished, their faces, legs, and abdomen often swell with water, making them appear well-fed to anyone who does not know the real cause of the bloating. (textbook, p. 117)

12. **b.** When problems occur in the nutrition of American toddlers, the cause is not usually inadequate availability of food, but family food practices that allow toddlers to eat when, what, and how much they choose. (textbook, p. 117)

13. **d.** Breast milk is more digestible than formula or cow's milk, which means that breast-fed babies have fewer allergies and digestive upsets than bottle-fed babies. (textbook, p. 115)

14. **d.** At birth, vision is the least developed of the senses. Newborns focus well only on objects that are about 10 inches away. (textbook, p. 109)

15. **a.** Risk of SIDS is higher for later-born males than for first-born males. (textbook, p. 106)

References

Bogin, B. (1988) *Patterns of growth.* Cambridge, England: Cambridge University Press.

Professor Bogin, who is heard on the audio program, presents a cross-cultural perspective on physical growth and discusses longitudinal studies of the effects of chronic malnourishment in early childhood on physical and intellectual development.

The First Two Years: Cognitive Development

ORIENTATION

During the first two years of life, cognitive development proceeds at a phenomenal pace as the infant is transformed from a baby who can know its world only through a limited set of basic reflexes into a toddler capable of imitating others, anticipating and remembering events, and pretending. Most significant among these advances is the development of language. By age 2, the average toddler has a relatively large vocabulary and is able to converse effectively with others.

Lesson 6 of *Seasons of Life* explores **cognitive development**—the ways in which individuals learn about, think about, and adapt to their surroundings— during the first two years. Chapter 6 of the textbook begins with a description and evaluation of Piaget's theory of **sensorimotor development.** From birth to age 2, infants learn about their environment by using their senses and motor skills. Although many psychologists believe that Piaget overemphasized motor skills and that development in most infants is more continuous than his stages imply, most accept his general outline of cognitive growth during infancy.

Audio program 6, "First Words," is concerned with language development from birth until the first word is spoken. As explained by psycholinguist Jill de Villiers, around the world babies and parents move along similar paths as language emerges. In the program, these paths are illustrated by actual examples and the description of linguistic landmarks such as **crying, cooing, babbling,** and the first true word. Along the way, the listener discovers the importance of the biological clock in the maturation of language, and the patterns of speech that adults use to promote linguistic development in their children. This issue—the interaction between maturation and learning in cognitive development—is explored further in the textbook, where several major theories of language development are evaluated, including those of B.F. Skinner and Noam Chomsky. Skinner argues that language development is the product of conditioning, while Chomsky maintains that children have a biological predisposition to acquire language. Most developmental psychologists nevertheless view it as an interactional process that reflects both **nature** (maturation) and **nurture** (conditioning).

As the program opens, we hear the voices of English- and Spanish-speaking parents reacting to a milestone in the lives of their children: their first word.

LESSON GOALS

By the end of this lesson you should be prepared to:

1. Outline and evaluate the theory of sensorimotor development proposed by Piaget.

2. Explain and contrast the theories of language development proposed by B.F. Skinner and Noam Chomsky.

3. Describe language development during the first two years.

Audio Assignment

Listen to the audio tape that accompanies Lesson 6: "First Words."

Write answers to the following questions. You may replay portions of the program if your memory needs to be refreshed. Answer guidelines may be found in the Lesson Guidelines section at the end of this chapter.

1. Outline the basic sequence and landmarks of language development from birth until the first word is spoken.

2. Identify the criteria used by developmental psychologists to determine whether an utterance represents the first actual word. What types of words are likely to be produced first?

3. Identify the characteristics of **baby talk,** the special form of language that adults use to talk to infants.

4. Explain how parents use **scaffolding** to encourage conversation in their children.

Textbook Assignment

Read Chapter 6, "The First Two Years: Cognitive Development," pages 121–143 in *The Developing Person Through the Life Span, 2/e.*

Write your answers to the following questions. Refer back to the textbook, if necessary. Answer guidelines may be found in the Lesson Guidelines section at the end of this chapter.

1. Describe the components of the six stages of sensorimotor development according to Piaget:
   Stage One:

   Stage Two:

   Stage Three:

   Stage Four:

   Stage Five:

   Stage Six:

2. Outline the basic sequence and landmarks of language development from the time the first word is spoken until the end of the second year.

3. Cite several criticisms of Piaget's theory of infant cognitive development.

4. Explain the disagreement between Piaget and many North American psychologists regarding the value of providing enriched learning programs for infants.

5. Compare and contrast the theories of B.F. Skinner and Noam Chomsky regarding early language development.

Testing Yourself

After you have completed the audio and text review questions, see how well you do on the following quiz. Correct answers, with text and audio references, may be found at the end of this chapter.

1. Translated literally, the word "infant" means:
   a. little scientist.
   b. not speaking.
   c. innocent one.
   d. explorer.

2. Between two and four months of age children begin making a pleasant, relaxing, speech sound called:
   a. babbling.
   b. cooing.
   c. scaffolding.
   d. baby talk.

3. At about six months of age babies begin to play with sounds in a manner that is more one of experimentation than socializing. This stage of language development is called:
   a. babbling.
   b. cooing.
   c. scaffolding.
   d. baby talk.

4. In order to qualify as a word, a sound must:
   a. serve as a symbol for something.
   b. be used consistently in a number of circumstances.
   c. resemble an adult word.
   d. possess all of the above characteristics.

5. When adults converse with children they usually do all of the following *except:*
   a. exaggerate their intonation.
   b. raise the pitch of their voices.
   c. use shorter sentences.
   d. speak only in the past tense.

6. During the sensorimotor period, babies:
   a. are incapable of thinking.
   b. "think" using their senses and movements.
   c. do not process sensory information.
   d. learn to manipulate simple symbols.

7. One objection to Piaget's theory of sensorimotor development is that he overemphasized the motor aspects of cognitive development. This criticism suggests that infants know:

    a. more than they are capable of communicating with their limited motor abilities.
    b. less than they are capable of communicating with their limited motor abilities.
    c. more today than they did when Piaget conducted his research.
    d. less than the highly gifted children Piaget studied.

8. Piaget believed that American psychologists and parents were too interested in:

    a. the social development of their children.
    b. accelerating cognitive development.
    c. the precise ages at which Piaget's stages were believed to occur.
    d. the physical development of their children.

9. The average baby says his or her first word at about:

    a. 6 months.
    b. 10 months.
    c. 12 months.
    d. 18 months.

10. For Noam Chomsky, the "language acquisition device" refers to the:

    a. deep structure of language.
    b. surface structure of language.
    c. human predisposition to learn language.
    d. portion of the human brain that processes speech.

11. According to Skinner, children acquire language:

    a. as a result of an inborn ability to use the basic structure of language.
    b. through reinforcement and learning by association.
    c. mostly because of biological maturation.
    d. in a fixed sequence of predictable stages.

12. When an infant understands that objects do not cease to exist when they are out of sight, he or she has attained the concept of:

    a. representation.
    b. displacement.
    c. object permanence.
    d. deferred imitation.

13. Children throughout the world attain similar language skills:

    a. during the second sensorimotor stage.
    b. during the fifth sensorimotor stage.
    c. in terms of surface structure but not deep structure.
    d. at about the same age.

14. Research on infant memory indicates that infants:
    a. cannot learn to remember images, but can remember behaviors.
    b. cannot learn to remember images or behaviors.
    c. can be taught to remember some images and behaviors earlier than had previously been thought.
    d. do not begin to develop retrievable memories until at least stage 3 of the sensorimotor period.

15. When a toddler uses a single word to convey a complete thought, he or she is using:
    a. baby talk.
    b. an overextension.
    c. a holophrase.
    d. deferred imitation.

LESSON 6 EXERCISE: BABY TALK

To further your understanding of the nature and significance of **baby talk,** make arrangements to listen to an adult conversing with an infant or toddler for ten to fifteen minutes. Your subjects may be family members, other relatives, or friends. The conversation need not be structured in any particular way. The adult might read to the child, play with a favorite toy, or simply carry on a conversation with the child. It is important that you not give the adult clues as to what speech patterns you are looking for. Ask your subject to relax, be candid, and enjoy interacting with the child. If you wish, you might even record the conversation to allow for a more thorough analysis later. After listening to the conversation, answer the questions on the **Exercise Response Sheet**. Return the sheet to your instructor.

LESSON GUIDELINES

Audio Questions Guidelines

1. Compared with that of other species, human hearing—even for subtle differences in speech sounds—is very sensitive at birth.

   Cries are the first speech productions on the way to words.

   Between two and four months, **cooing** begins. Cooing is a pleasant and relaxing sound with an important social function: it attracts the attention of caregivers.

   At about six months infants begin to mix consonants with cooing. As they play with sound, they produce utterances such as "ba, ba, ba, ba," and enter the stage of **babbling.** Babbling is not as social as cooing, but represents experimentation with the sounds of language.

   Infants below the age of about eight months babble in very similar ways, so that the babbling of children from around the world is indistinguishable. At about ten months babbling begins to take on the characteristics of the language the child will ultimately learn.

   The first true word usually comes at about 12 months of age, although there is wide, and normal, variation in the age.

2. The first criterion is that the sound must be a symbol referring to an object or an event.

   Secondly, the sound must be used consistently in a number of different circumstances.

   Thirdly, the sound must resemble an adult word.

   The first words are likely to refer to something that moves, something the baby controls, and something that is very interesting. Examples of such words are toys, foods, clothing, mothers, fathers, siblings, and pets.

3. **Baby talk** differs from adult speech in several ways: it is distinct in pitch (higher), intonation (more exaggerated and more low-to-high fluctuations), vocabulary (simpler and more concrete), and sentence length (shorter).

   Baby talk is also more repetitive and uses more questions, fewer past tenses, pronouns, and complex sentences.

   An intriguing fact is that people of all ages, including nonparents, use baby talk when conversing with infants.

   In many of the world's cultures, baby words for mother and father—such as "mama," and "dada"—are just the sort of easy-to-produce, repeated syllables that often are a child's first words.

4. **Scaffolding** refers to the tendency of parents and other adults to try to support the young child's conversation. In reading a book, for example, the parent might first read the entire story slowly and with exaggerated intonation to give the child time to process the information. The second time the story is read, the parent might pause at key points and wait for the child to respond in some way. Gradually, the parent removes more and more of the

support, or scaffolding, so that the child's role in the conversation becomes more extensive, and precise.

Textbook Question Guidelines

1. Stage 1: (birth–1 month) The newborn's reflexes, such as sucking, grasping, looking, and listening, represent its only **schemas,** or ways of learning about the world.

   Stage 2: (1–4 months) The infant begins to adapt his or her reflexes to the environment and to coordinate actions (e.g., grabbing a bottle and then sucking from it).

   Stage 3: (4–8 months) Infants become more responsive to people and objects in their environments.

   Stage 4: (8–12 months) Infants become more purposeful in responding to people and objects, anticipating events and reaching for desired objects. **Object permanence** is established.

   Stage 5: (12–18 months) The **little scientist** becomes more active and creative in the trial-and-error exploration of his or her environment.

   Stage 6: (18–24 months) By using **mental combinations** toddlers begin to solve simple problems without needing to use trial-and-error experimentation.

2. Around the time of the first birthday, the first words that are recognizably part of the native language are spoken.

   Between 12 and 18 months of age, vocabulary slowly grows to about 50 words.

   At first, infants **overextend** the few words they know to a variety of inappropriate contexts, indicating that language development proceeds much as did cognitive development, with the child forming and testing hypotheses.

   A principle of language development is that at each stage children seem to understand much more than they are capable of expressing, as is indicated by the child's use of a single word, or **holophrase,** to express a complete thought.

   The first two-word sentence usually appears between 18 and 21 months.

   By 24 months, the toddler has a vocabulary of more than 200 words. Grammar is apparent in word order, suffixes, prefixes, and use of pronouns.

3. Many developmental psychologists believe that the typical child's cognitive development is much more gradual and continuous than is implied by Piaget's delineation of distinct mental stages.

   Several researchers have questioned Piaget's methods of measuring cognitive development, arguing that the tests can be influenced by factors that Piaget failed to consider.

   Some critics believe that Piaget's theory places too much emphasis on the motor, rather than the sensory, aspects of infants' cognitive development. Many believe that as a result, he underestimated the perceptual abilities present during the first six months of life.

4. Piaget viewed infants as actively organizing their own experiences, and made little mention of the role of specific learning experiences. This reflects his view that development should not be rushed.

   Many American psychologists believe that society can, and should, foster cognitive development in children by providing appropriate instruction and enriching experiences.

   Recent research has shown that infants can be taught many skills at an early age.

   Most psychologists would agree, however, that too much acceleration and stimulation can be detrimental.

5. B.F. Skinner maintains that verbal behavior is acquired through conditioning, learning by association, and the differential reinforcement of appropriate vocalizations.

   The structural view espoused by Noam Chomsky maintains that children have an innate predisposition (**language acquisition device**) to learn language. According to this view, children have an inborn understanding of the basic, or **deep structure**, of language. What they must learn is the vocabulary and grammar, or **surface structure**, of language.

   Research on early language development indicates that language development results from the *interaction* of maturational processes and environmental experiences.

Answers to Testing Yourself

1. **b.** Translated from its Latin roots, the word "infant" refers to someone who is "not yet speaking." (audio program)

2. **b.** Cooing serves the very important social function of attracting the attention of the caregiver. (audio program; textbook, p. 137)

3. **a.** Babbling is sheer experimentation with the sounds of language. (audio program; textbook, p. 137)

4. **d.** To qualify as an actual word, sounds must be symbols, resemble adult words, and be used consistently. (audio program)

5. **d.** In speaking to children, adults generally confine their conversation to things in the "here and now." (audio program)

6. **b.** Piaget's insight into cognitive development was that babies, unlike adults, think exclusively with their senses and motor skills. (textbook, p. 122)

7. **a.** Most researchers now agree that Piaget underestimated early perceptual abilities and hence certain aspects of cognitive development during the first six months of life. (textbook, p. 131)

8. **b.** In fact, Piaget referred to "What should we do to foster cognitive development?" as the "American question." (textbook, p. 132)

9. **c.** At about 12 months, the first words that are recognizably part of the child's native language are spoken. (audio program; textbook, p. 137)

10. **c.** The LAD emphasizes the inborn, automatic nature of the infant's learning ability. (textbook, p. 137)

11. **b.** According to learning theorists, the words and phrases children learn depend on the nature, frequency, and timing of the conditioning process. (textbook, pp. 135–137)

12. **c.** Object permanence is the major intellectual accomplishment that signals the beginning of Stage 4 in Piaget's theory. (textbook, p. 124)

13. **d.** Children all over the world follow the same sequence and approximately the same timetable for early language development. (textbook, p. 137)

14. **c.** Recent studies have shown that infants as young as 8 weeks old are able to retain simple learned movements after a long interval (18 days). (textbook, p. 133)

15. **c.** Holophrases demonstrate that even a single word, amplified by intonation and gestures, can express a whole thought. (textbook, p. 139)

References

de Villiers, Jill G., and de Villiers, Peter A. (1978). *Language acquisition.* Cambridge, MA: Harvard University Press.

A scholarly description of early language development by two eminent researchers.

The First Two Years: Psychosocial Development

AUDIO PROGRAM: Attachment: The Dance Begins

ORIENTATION

Lessons 5 and 6 of *Seasons of Life* examined physical and cognitive development during the first two years. Lesson 7, which is concerned with psychosocial development during infancy, explores the individual's emerging self-awareness, personality, emotional expression, and relationship to parents and society.

As discussed in Chapter 7 of *The Developing Person Through the Life Span, 2/e,* contemporary developmentalists have revised a number of the traditional views of psychosocial development. It was once believed, for example, that infants did not have any real emotions. Researchers now know, however, that in the very first days and weeks of life, infants express and sense many emotions, including fear, anger, happiness, and surprise.

In the traditional view of personality development, the infant was seen as a passive recipient of the personality created almost entirely by its mother's influence. But it is now apparent that many personality traits are present in infants at birth, before parental influence is felt. In addition, active parent–infant interaction within a secure and nurturing environment is now viewed as a central factor in the child's psychosocial development.

After treating these new perspectives, the text goes on to discuss **child abuse** and neglect. Parental maltreatment is explained as the product of many interrelated factors, including the parents' background, the child's behavior, the stresses of daily life, and even cultural attitudes towards children.

Audio program 7, "Attachment: The Dance Begins," explores the who, when, where and why of **attachment,** the affectional tie between infants and their primary caretakers. In infants the world over, attachment develops at about 7 months—when babies start to be aware that other people stay in existence even when they're out of sight. Attachment helps ensure that the relatively helpless human infant receives the adult care it needs in order to survive.

But what about infants who do not become securely attached? And what are the effects, for example, of adoption and day care on the development of attachment? Through the expert commentary of psychologists Michael Lamb, Janice Gibson, and Sheldon White, we explore these important issues in attachment—an intricate "interaction" between infant and caregiver that lays the foundation for psychosocial development throughout the subsequent seasons of life.

LESSON GOALS

By the end of this lesson you should be prepared to:

1. Outline the development of basic emotions and personality characteristics, both in theory and as observed in fact.

2. Discuss the significance of parent–infant interaction in the infant's psychosocial development.

3. Describe the characteristics of a good home environment, a poor home environment, and an abusive one.

4. Discuss issues related to attachment, including its timing, participants, and long-term impact on psychosocial development.

Audio Assignment

Listen to the audio tape that accompanies Lesson 7: "Attachment: The Dance Begins."

Write answers to the following questions. You may replay portions of the program if you need to refresh your memory. Answer guidelines may be found in the Lesson Guidelines section at the end of this chapter.

1. Define attachment, outline its development, and explain why experts believe it to be a biologically determined event.

2. Discuss the impact of early mother-infant contact, adoption, and day care on attachment.

3. Discuss the immediate and long-range impact on infants of secure and insecure attachment.

Textbook Assignment

Read Chapter 7: "The First Two Years: Psychosocial Development," pages 145–174 in *The Developing Person Through The Life Span, 2/e.*

Write answers to the following questions. Refer back to the textbook, if necessary. Answer guidelines may be found in the Lesson Guidelines section at the end of this chapter.

1. Identify and describe the developmental course of at least three emotions that are shown during the first two years of life.

2. Identify and describe the first two stages of personality development, according to Freud's theory.

3. Identify the first two stages of personality development according to Erikson's theory, and describe the healthy outcome of each.

4. Describe the stages of infant and toddler personality development proposed by Mahler and their implications for adult personality.

5. Discuss the significance of synchrony and attachment in parent–infant interaction and psychosocial development.

6. Identify the characteristics of a child's home environment that promote psychosocial development.

7. Outline the extent of child abuse and neglect, and describe how it might be treated or prevented.

Testing Yourself

After you have completed the audio and text review questions, see how well you do on the following quiz. Correct answers, with text and audio references, may be found at the end of this chapter.

1. Attachment between infant and primary caregiver typically happens at about what age?
   a. One or two months
   b. Three or four months
   c. Six or seven months
   d. Ten or eleven months

2. Experts believe that the emergence of attachment in infants is:
   a. biologically based and linked to the cognitive ability to represent other individuals.
   b. a learned behavior elicited by responsible caregiving.
   c. largely dependent on early bonding between the mother and infant.
   d. unpredictable and of relative unimportance in later psychosocial development.

3. Infants placed for adoption after six months of age:
   a. never form secure attachments.
   b. can still become securely attached to adoptive parents.
   c. are more likely to be abused than infants adopted at an earlier age.
   d. gain weight more slowly than infants adopted at an earlier age.

4. Most infants who are regularly placed in day care:
   a. develop attachment to day-care workers rather than parents.
   b. become insecurely attached, because of their confusion regarding their primary caregivers.
   c. show signs of extreme anxiety when tested in the strange person situation.
   d. form secure attachments to their parents.

5. Stuffed animals, favorite blankets, and other objects that children keep near them when apart from their primary caregivers are called:
   a. security objects.
   b. transitional objects.
   c. vicarious parents.
   d. symbolic pacifiers.

6. One emotion that appears in early infancy is:
   a. anger.
   b. joy.
   c. fear.
   d. shame.

7. In Erik Erikson's theory, the crisis of infancy is one of:
    a. trust versus mistrust.
    b. shame versus doubt.
    c. autonomy versus dependence.
    d. industry versus inferiority.

8. Parents who abuse their children are most likely to be:
    a. older and unintelligent.
    b. older and reclusive.
    c. younger and poorly educated.
    d. There are no predictable traits of abusive parents.

9. Compared to mothers, fathers are more likely to:
    a. engage in physical play.
    b. encourage achievement motivation.
    c. encourage intellectual development.
    d. encourage social development.

10. An important effect of secure attachment is the promotion of:
    a. egocentrism.
    b. curiosity and self-directed behavior.
    c. dependency.
    d. all of the above.

11. Of the following personality traits, which was *not* identified as remaining stable as children get older?
    a. Activity level
    b. Sociability
    c. Fearfulness
    d. Quality of mood

12. Infants who show a fear of strangers:
    a. are most likely insecurely attached.
    b. were probably placed in day care too early.
    c. are most likely to be securely attached.
    d. are displaying a universal emotion.

13. A child at Erikson's autonomy stage would also be at Freud's:
    a. oral stage.
    b. anal stage.
    c. phallic stage.
    d. latent stage.

14. The carefully coordinated play of infants and their parents is described as:
    a. synergy.
    b. symbiosis.
    c. synchrony.
    d. duetting.

15. The HOME scale is designed to predict:
    a. physical development during the preschool years.
    b. cognitive development in children.
    c. which children are likely to be the victims of neglect or abuse.
    d. whether an infant will become securely or insecurely attached.

LESSON 7 EXERCISE: ATTACHMENT AND THE "STRANGE SITUATION"

About 7 months after birth—around the time when children develop the ability to represent another person cognitively—infants develop an enduring affectional **attachment** to their primary caregivers.

Attachment can be measured in many ways. Infants express attachment by "proximity-seeking" behaviors, such as approaching, following, and clinging; and "contact-seeking" behaviors, such as crying, smiling, and calling. Parents express their attachment more by eye contact than by physical contact, and by reacting to their child's vocalizations, expressions, and gestures.

On the basis of many naturalistic observations, Mary Ainsworth developed a laboratory procedure in which the infant's reactions to a novel situation and the comings and goings of its caregiver indicate the security of the child's attachment. In this test, which is conducted in a well-equipped playroom full of toys, most infants demonstrate **secure attachment.** The presence of their mother gives them the sense of security needed to express their natural curiosity and explore the new room. If their mother attempts to leave the room, securely attached infants will usually stop playing, protest verbally, and demonstrate contact-seeking behaviors.

Approximately one-third of infants show insecure attachment in this test situation, clinging nervously to their mother and being unwilling to explore even while she remains in the room. Others seem aloof and engage in little or no interaction with their mothers.

To better understand how attachment is measured, arrange to observe a one- or two-year-old and his or her caregiver in a play setting outside of the child's home. Ideally, ask a relative or friend and their child to participate. The play setting could be in your home, at a local playground, or any other mutually agreeable location. If you do not know someone with a young child, you can complete this exercise by visiting a playground or day care center.

Before your scheduled observation period, read through the questions on the **Exercise Response Sheet** so that you will know what behaviors to watch for. Observe your participants for ten to fifteen minutes of unstructured play. If possible, during the observation period, ask the adult to make a move as if he or she were going to leave. Observe the child's reaction. After the observation period, complete the following questions and return the response sheet to your instructor.

LESSON GUIDELINES

Audio Question Guidelines

1. **Attachment** refers to the process by which infants develop a lasting affectional tie with their primary caregiver. Attachment goes two ways, from parent to child and from child to parent.

   From the parent's point of view, bonding begins during pregnancy, includes the special memories of birth, and continues to develop indefinitely.

   Babies give their first **social smiles** at about 4 to 6 weeks of age.

   Attachment emerges in children the world over at about 7 months of age. This regularity implies that it is a biologically based event.

   Attachment ensures that a newborn as immature and helpless as the human will receive the adult help it needs in order to survive.

   The signs of attachment to a particular person are clear. The baby turns to that person when distressed and protests when that person leaves. Babies may also cry when strangers appear.

2. There is no good evidence that early mother–infant contact has a major impact on attachment. There is no reduction in the quality of the relationship formed between mothers and babies when contact in the first few weeks of life does not take place.

   Because babies form their first attachment at about 7 months of age, it is somewhat easier to place a child for adoption before that time. A child placed with adoptive parents after that age may have to go through a process of grieving the loss of an earlier attachment figure before investing emotionally in a new one.

   In most circumstances, babies form more than one attachment in the course of growing up.

   Approximately two-thirds of babies in this country form secure attachments to their mothers whether or not they are in regular day care. An important variable in day care is the quality of care that the child receives and the extent to which it matches the style of the parents.

   Because their children may develop attachments to day-care workers, day care may be a more difficult adjustment for parents than it is for children.

3. Three-year-old children who were **securely attached** at one year are more mature and significantly more independent, self-confident, cooperative, and sociable, than those who were insecurely attached.

   Securely attached children are more likely to be persistent and resilient in challenging situations.

   The developmental advantages of securely attached children seem to continue to age 5 or 6, when children start school. Most likely, the reason is that their home has been consistently nurturant.

Textbook Question Guidelines

1. Although it was once thought that infants did not have real emotions, researchers have discovered that even very young infants express and respond to many emotions.

   When infants are only a few hours old, they cry, look surprised, and show other signs of fear in response to a loud noise or a sudden loss of support.

   Sensitivity to sadness is apparent early in infancy.

   Pleasure smiles appear in the first days of life.

   Social smiles begin to appear at about 4 to 6 weeks.

   Between 8 months and 2 years, infants' emotions become stronger, more varied, and distinct.

   Fear of strangers is first noticeable at about 6 months and reaches its maximum at 12 months.

   Separation anxiety appears at 8–9 months, peaks at about 14 months, and then gradually diminishes as the second birthday approaches.

   Anger becomes more common during toddlerhood.

2. According to Freud, infants in the first year are in the **oral stage** of psychosexual development, in which the mouth is the primary source of gratification and pleasure.

   According to this view, the mother's attitudes toward feeding and weaning the infant are a critical factor in the infant's psychological development.

   If infants become "fixated" in the oral stage, as adults they may engage in excessive eating, drinking, smoking, or talking, in an effort to obtain the oral gratification they were denied in infancy.

   During the second year (**anal stage**) pleasure is derived from stimulation of the bowels. The child's behavior changes from the passive and dependent style of the oral stage to the expulsive and more independent mode of the anal stage.

   Overly strict or premature toilet training will produce adults with "anal" personalities who overemphasize cleanliness and precision (retentive), or are messy and disorganized (expulsive).

3. Erikson's theory focuses on the overall patterns of child-rearing during the first two years.

   The basic conflict of infancy is one of **trust versus mistrust.** Trust develops in babies if they are kept well-fed, warm, and comfortable. Trust promotes the development of a secure sense of self.

   The conflict of toddlerhood is that of **autonomy versus shame and doubt.**

   For many toddlers, the struggle for autonomy centers around toilet training.

   Infants who fail to develop trust or achieve autonomy may develop into self-doubting, suspicious, and pessimistic adults.

4. **Normal autism:** From birth until about 2 months, infants are self-absorbed and dominated by the primitive needs for food and sleep.

   Symbiosis: From 2 months to 5 months, the infant's dependence on the mother is so complete that the infant feels physically a part of the mother. This dependence provides a secure base for later independence.

   Separation–individuation: From 5 months on, the infant gradually develops a sense of self separate from the mother.

   According to Mahler, maladaptive mothering during infancy has irrevocable implications, producing adults who fear independence, avoid intimacy, and fail to develop a mature sense of self.

5. Using frame-by-frame analysis of filmed parent–infant interaction, researchers have found an impressive **synchrony** between parent and child, with each partner giving vocal and gestural cues to the other.

   The absence of synchrony, as when parents are insensitive to the infant's signals, or misinterpret the infant's temperament, can also influence the child's psychosocial development.

   Attachment to a familiar person grows stronger over the first 2 years.

   Secure attachment, in which the infant shows signs of missing the mother when she leaves and welcoming her when she returns, is one indication of the quality of care in early infancy.

   Insecure attachment is evident in infants who are angry and inconsolable when their mothers leave, who show no distress when their mothers leave, or who avoid their mothers.

   Secure attachment aids cognitive development and predicts future preschool achievement. Securely attached infants are more curious, outgoing, and self-directed.

6. Cognitive development during toddlerhood is fostered in infants who have responsive, involved parents and a safe, stimulating play environment.

   One method of measuring environmental quality is **HOME:** the Home Observation for the Measurement of the Environment. The six subscales of HOME include emotional and verbal responsiveness of the mother, avoidance of restriction and punishment, organization of the physical environment, provision of appropriate play materials, maternal involvement with the child, and opportunities for variety in daily stimulation.

7. In the United States, 1 out of every 43 children under age 14 was reported as abused or neglected in 1982.

   Child abuse includes physical abuse (battered child syndrome), sexual abuse, emotional maltreatment, and **neglect.**

   Many childhood accidents (the most common cause of childhood death and injury) can be traced to neglect.

   In cultures that have a low rate of child abuse, children are more highly valued, child care is shared by a greater number of people, and young children are not expected to be responsible for their actions, which results in less use of punishment.

Childhood abuse and neglect are the product of many interrelated factors, including variables related to personality, social setting, and background.

Since youth, poverty, and ignorance correlate with parental maltreatment, the prevention of child abuse will probably depend on measures to raise the lowest income levels, discourage teenage parenthood, and increase parents' levels of education.

Answers to Testing Yourself

1. **c.** In children throughout the world, attachment emerges at 6 or 7 months. (audio program)

2. **a.** Attachment emerges at about the same age that infants develop the cognitive ability to maintain the image of another person in their minds. (audio program)

3. **b.** Infants who are placed for adoption after the age of six months may first grieve the loss of an earlier attachment, but they can still develop new secure attachments. (audio program)

4. **d.** Infants placed regularly in day care may also become attached to day care workers, but they still become attached to their parents. (audio program)

5. **b.** Transitional objects help fill an emotional need in the time between when a child is physically near the parent and when he or she can be away from the parent completely. (audio program)

6. **c.** Fear is one of the first emotions that can be discerned in infants. (textbook, p. 145)

7. **a.** Babies learn to trust their world if they are kept well-fed, warm, and dry significantly more often than they are left hungry, cold, and wet. (textbook, p. 150)

8. **c.** Abusive parents often are under age 20, have little education, and were abused or neglected themselves as children. (textbook, p. 172)

9. **a.** Fathers typically are noisier, more physical, and make bigger gestures in their play than mothers do. (textbook, p. 166)

10. **b.** Children who were securely attached infants are more curious, competent, and self-directed than those who were insecurely attached. (textbook, p. 162)

11. **d.** Quality and rhythmicity of mood are quite variable throughout life. (textbook, p. 154)

12. **d.** Fear of strangers is universal. (textbook, p. 146)

13. **b.** For many toddlers, the struggle for autonomy centers around toilet training. (textbook, pp. 150–151)

14. **c.** Synchrony refers to the intricate meshing of the behaviors of caregivers and their babies. (textbook, p. 158)

15. **b.** HOME (Home Observation for the Measurement of the Environment) is a list of 45 family and household characteristics that have been shown to correlate with children's cognitive development. (textbook, p. 163)

References

Sroufe, L. Alan. (1978). Attachment and the roots of competence. *Human Nature, 1,* 50–57.

Sroufe, L. Alan. (1985). Attachment classification from the perspective of infant–caregiver relationships and infant temperament. *Child Development, 56,* 1–14.

Professor Sroufe discusses the importance of attachment in psychosocial development.

The Play Years: Physical Development

ORIENTATION

Lesson 8 is the first of a three-lesson unit that describes the developing person from 2 to 6 years in terms of physical, cognitive, and psychosocial development. Lesson 8 examines physical development during the play years.

Children grow steadily taller and slimmer during the preschool years, with their genetic background and nutrition being responsible for most of the variation seen in children from various parts of the world. The most significant aspect of growth is the continued maturation of the nervous system and the refinement of the visual, muscular, and cognitive skills that will be necessary for the child to function in school. The brain also becomes more specialized, with the left side usually becoming the center for speech and the right the center for visual, spatial, and artistic skills.

One of the themes of the textbook is that "play is the work of children." Chapter 8 includes descriptions of **sensorimotor play, mastery play,** and **rough-and-tumble play,** and delivers the important message that through play children acquire the motor and social skills they will need to function as adults.

Play is especially appropriate for the extended period of childhood that has been programmed into the biological clock. Audio program 8, " 'How To' Time," explores the evolutionary origins of this period and the use to which humans have put it.

The human biological clock delays the onset of reproductive maturity and gives children time to acquire what Erik Erikson referred to as a sense of **industry.** Childhood is the time to pretend, to play, and to learn how to do things. And herein lies the key to our extended childhood. More than any other species, we humans are dependent on complex, learned behavior for our survival. Through the expert commentary of pediatrician Howard Weinblatt, endocrinologist Inese Beitins, anthropologist Barry Bogin, and evolutionary biologist Stephen Jay Gould, the significance of "how to" time is explored.

The program opens with the voice of a girl explaining how to ride a tricycle, a skill rarely used by adults but never forgotten.

LESSON GOALS

By the end of this lesson you should be prepared to:

1. Describe physical development during the preschool years and discuss how physical maturation is related to school readiness.

2. Outline the development of gross and fine motor skills during early childhood, and describe activities that foster these skills.

3. Compare and contrast physical development in preschool girls and boys.

4. Discuss the significance of the extended period of human childhood and the importance of play in development.

Audio Assignment

Listen to the audio tape that accompanies Lesson 8: "'How To' Time."

Write answers to the following questions. You may replay portions of the program if you need to refresh your memory. Answer guidelines may be found in the Lesson Guidelines section at the end of this chapter.

1. Outline the biological clock's hormonal program for timing sexual maturity.

2. Cite two possible evolutionary explanations of why humans have such a lengthy period of childhood.

3. Explain the "down-and-out" principle of physical development.

Textbook Assignment

Read Chapter 8: "The Play Years: Physical Development," pages 179–195 in *The Developing Person Through the Life Span, 2/e.*

Write your answers to the following questions. Refer back to the textbook, if necessary. Answer guidelines may be found in the Lesson Guidelines section at the end of this chapter.

1. Outline normal physical growth during the play years, and account for variations in growth.

2. Outline the relationships between brain specialization, language development, and handedness.

3. Cite and discuss two arguments in the controversy concerning when children should learn to read.

4. List three factors that influence a child's chances of having an accident.

5. Differentiate between gross and fine motor skills and describe their development during the play years.

6. Identify three kinds of play and the skills they develop.

7. Outline the main similarities and differences in the physical development of boys and girls during the play years.

Testing Yourself

After you have completed the audio and text review questions, see how well you do on the following quiz. Correct answers, with text and audio references, may be found at the end of this chapter.

1. The primary sex hormones are ―――――― in females and ――――――
   in males.
   a. testosterone . . . estrogen
   b. estrogen . . . testosterone
   c. adrenaline . . . noradrenaline
   d. progesterone . . . testosterone

2. Which of the following most accurately describes how the levels of sex
   hormones change from birth to puberty?
   a. In both males and females, levels of sex hormones increase from birth
      until puberty.
   b. In males only, levels of sex hormones increase from birth until puberty.
   c. In females only, levels of sex hormones increase from birth until
      puberty.
   d. In both males and females, levels of sex hormones are very high at birth,
      drop at about 18 months, and become high again at puberty.

3. Anthropologists have proposed that an extended period of childhood
   evolved in humans because:
   a. humans need this time to learn the many complex behaviors critical for
      survival.
   b. babysitting by older children conferred such an advantage to humans
      that our biological clock evolved to keep childhood as lengthy as possi-
      ble.
   c. both a and b are true.
   d. none of the above are true.

4. The "down-and-out" principle of development would explain why:
   a. reproductive maturity does not occur until puberty.
   b. babies can sit up before their color vision is mature.
   c. babies can sit up before they can stand.
   d. physical development is more rapid in girls than in boys.

5. According to developmental psychologists, the play of "how to" time is an
   important way in which children:
   a. perfect physical skills.
   b. develop a sense of being useful and competent.
   c. develop what Erikson referred to as a sense of industry.
   d. do all of the above.

6. During the preschool years, the most common nutritional problem is:
   a. malnutrition.
   b. iron deficiency anemia.
   c. the junk food syndrome.
   d. vitamin B deficiency.

7. As the brain matures, the right side becomes specialized for _____ skills and the left for _____ skills.
   a. language . . . spatial
   b. artistic . . . speech
   c. logical . . . musical
   d. analytical . . . synthetic

8. Signs of left- or right-hand preference become apparent:
   a. during infancy.
   b. when children are about 1.
   c. at about 2 years of age.
   d. when children start school.

9. Throughout childhood, the leading cause of death is:
   a. suffocation.
   b. congenital defects.
   c. disease.
   d. accidents.

10. The infant who delights in watching a turning mobile is engaging in:
    a. sensorimotor play.
    b. mastery play.
    c. rough-and-tumble play.
    d. too little information to determine.

11. Concerning sex differences in childhood, which of the following is true?
    a. Knowing a child's sex tells nothing about development.
    b. Girls generally are better than boys at gross motor skills.
    c. Girls generally are better than boys at some fine motor skills.
    d. In girls, the right side of the brain develops more rapidly than the left side.

12. During each of the preschool years, children grow approximately _____ inches and gain approximately _____ pounds.
    a. 1 . . . 3
    b. 2 . . . 7
    c. 3 . . . 4.5
    d. 4 . . . 10

13. The child's ability to do elementary-school work is most directly related to:
    a. social maturation.
    b. brain maturation.
    c. development of the visual system.
    d. fine motor skill development.

14. Most differences in strength and skill development during the play years are the result of:

    a. sex differences.
    b. differences in how children are socialized.
    c. inherited abilities.
    d. nutritional differences.

15. During early childhood, the most rapidly developing part of the body is the:

    a. heart.
    b. reproductive system.
    c. visual system.
    d. brain.

LESSON 8 EXERCISE: PLAY SPACES AND PLAYGROUNDS

A major theme of this lesson is that "play is the work of childhood." According to most developmental psychologists, both work and play are important activities at every stage of life. But the line between play and work is not clear-cut, especially in the play years. Rather than thinking about the two as opposites, many developmentalists argue that it is more profitable to think of them as endpoints along a continuum, from the most whimsical, spontaneous, nonproductive play on the one side, to the most deliberate, planned, productive work on the other.

Developmental psychologists view play as the major means through which physical, cognitive, and social skills are mastered. Unfortunately, many adults are so imbued with the work ethic that they tend to denigrate children's play. Some even punish their children for "horsing around," criticize nursery school teachers for letting children play "too much," or schedule their children's lives so heavily with lessons, homework, and chores that there is little time left for play.

Every age has its own special forms of play. Play that captures the pleasures of using the senses and motor abilities is called **sensorimotor play.** The major play of infants, this type of play continues throughout childhood. Much of the physical play of childhood is **mastery play,** which refers to play that helps children to master new skills. This type of play is most obvious when physical skills are involved, but increasingly comes to include intellectual skills as children grow older. A third type of play is the **rough-and-tumble** wrestling, chasing, and hitting that occur purely in fun, with no intent to harm.

Young children need safe, adequate play space and the opportunity to play with children their own age. To increase your awareness of the play and play needs of children in your neighborhood, the exercise for Lesson 8 asks you to locate and observe play spaces within walking distance of your home. If there are none, what alternatives exist for preschoolers in your neighborhood? For example, do they play in backyards? Or are most preschoolers enrolled in nursery schools that have their own play spaces? Please answer the following questions and return your completed **Exercise Response Sheet** to your instructor.

LESSON GUIDELINES

Audio Question Guidelines

1. The biological clock is a metaphor for the body's mechanisms for timing physical development.

   Hormones are chemical messengers secreted into the bloodstream by glands and have a variety of effects on the body. **Sex hormones,** such as **estrogen** in females and **testosterone** in males, are responsible for triggering development of the reproductive system.

   By the end of the third month of prenatal development, the human fetus has sex organs and is producing sex hormones. At birth, sex hormones are present in a baby's bloodstream at levels equivalent to those at puberty.

   At about 18 months of age, the high levels of sex hormones drop and remain at a very low level until puberty, when there is another surge and reproductive maturity is attained.

2. Anthropologists have suggested that one reason humans have a much longer childhood than other animals (and that the development of their reproductive system is delayed) is to provide an opportunity for children to learn the many complex behaviors that give our species its great cognitive capacity. This explanation is supported by the fact that our species' survival depends to a much greater extent on learned behavior than does that of any other species.

   Another possible explanation is that an extended childhood allows older children to take care of younger children. In hunting and gathering societies throughout the world, the prime job of children is taking care of younger children. In animals, however, the mother must care for the young herself and is therefore limited in her capacity to produce more offspring.

3. The "down-and-out" principle of development is that maturation begins with the head and brain and works its way down the spine and out to the extremities. In the textbook, these principles are referred to as cephalo-caudal (head to tail) and proximo-distal (near to far, or inside out) development.

   This principle explains why babies are able to lift their heads before they can sit up and can sit before they are able to stand. It also explains the normal progression of gross motor skill development. For example, younger children throw balls with their entire arms. Older children are able to control a throw more precisely by adding, successively, motion of the wrist, hand, and fingers.

Textbook Question Guidelines

1. Between 2 and 6 years of age children grow almost 3 inches and gain about 4.5 pounds.

   By age 5, the average North American child weighs about 40 pounds and measures 43 inches.

   Height and body weight show great variation from child to child, with nutrition and genetic background being the greatest influences on these differences.

2. During childhood, the brain develops faster than any other part of the body, attaining about 90 percent of its adult weight by age 5. **Myelination,** which insulates nerves and speeds neural transmission, accounts for much of this brain growth.

   Many abilities, including skills needed for school readiness such as reading and writing, are dependent on a certain level of myelination.

   Brain areas associated with eye-hand coordination become fully myelinated about age 4; those associated with attention, not until the end of childhood; those associated with language and intelligence, not until age 15 or so.

   The two **hemispheres** of the brain become more specialized as the brain matures, with the left side specializing in logical analysis and language, and the right side in visual and artistic skills, facial recognition, spatial perception, and responses to music. These relationships hold for about 95 percent of right-handed adults, and 70 percent of left-handed adults.

   As the **corpus callosum** becomes increasingly myelinated between ages 2 and 8, the two sides of the brain become more closely integrated.

3. Some experts advise parents to begin teaching school skills early, in order to make their child smarter and less likely to be frustrated once he or she begins school.

   Others contend that many children, especially boys, are not ready for formal teaching until age 8–10. They contend that children who are forced to begin schooling too early are more likely to become frustrated and achieve less.

   Most developmentalists agree that readiness to read depends on the individual child's maturation, interests, and experiences, not on any chronological age.

4. Activity level is higher during the early preschool years (2–3 years) than at any other time in the life span.

   Throughout childhood, accidents are the leading cause of death.

   A child's chances of having an accident depend primarily on three factors: the amount of adult supervision, the safety of the play space, and the child's activity level.

   Groups of children who have more accidents than others are boys, impoverished children, and black children.

5. **Gross motor skills** (large body movements such as running, climbing, jumping, and throwing) improve dramatically between ages 2 and 6.

   Fine motor skills involving small body movements (pouring, using knife and fork, etc.) are much harder for preschoolers to master. This difficulty is due to several factors, including incomplete myelination of the nervous system, incomplete muscular control, and having short, fat fingers that are unsuited for many utensils, toys, and clothes.

   Many educators, including Maria Montessori, consider development of fine motor skills to be an important part of an enriching preschool curriculum.

6. **Sensorimotor play** captures the pleasures of sensory experiences and motor abilities.

   Mastery play describes play that helps the child to master new skills, such as throwing a ball, skipping, or using scissors.

   Rough-and-tumble play refers to physical play in which aggression is feigned, often with the participants exhibiting a **play face.**

   Rough-and-tumble play is a social activity that usually occurs among children who have considerable social experience. It is slow to develop in younger children, newcomers, and only children.

7. Boys and girls follow nearly identical paths of physical development during early childhood. Variation in most skills is greater among children within either sex than between the sexes.

   Boys are slightly taller, more muscular, and have less body fat than girls. By age 4, boys are usually superior in activities that require gross motor skills involving limb strength and control.

   Girls' bone age is usually at least one year ahead of boys of the same age. Girls are more coordinated in skipping, kicking, and in certain fine motor skills, especially those involving speed.

   Male-female differences may be the result of different amounts of practice on certain skills, rather than intrinsic body or brain differences.

Answers to Testing Yourself

1. **b.** Estrogen is the primary female sex hormone, testosterone the primary male hormone. (audio program)

2. **d.** At birth, the level of sex hormones in both sexes is as high as it will be at puberty. In the years in between, however, levels drop precipitously. (audio program)

3. **c.** Anthropologists believe that the benefits of learning and babysitting resulted in the evolution of an extended childhood in humans. (audio program)

4. **c.** The muscles that permit sitting develop before those that permit standing, following the head-down principle of development. (audio program)

5. **d.** Play is the "work" of childhood. (audio program)

6. **b.** Iron deficiency anemia stems from insufficient intake of quality meats and dark-green vegetables. (textbook, p. 181)

7. **b.** The left side of the brain becomes specialized for speech and logical analysis; the right side for responses to music, facial recognition, artistic skills, and spatial perception. (textbook, p. 182)

8. **a.** Some signs of hand preference are evident in early infancy. (textbook, p. 183)

9. **d.** Accidents are the leading cause of death among children from ages 1 to 11. (textbook, p. 186)

10. **a.** Sensorimotor play captures the pleasures of using the senses and motor abilities. (textbook, p. 189)

11. **c.** Especially when speed is involved, girls tend to be better than boys at certain fine motor skills. (textbook, p. 193)

12. **c.** (textbook, p. 179)

13. **b.** Brain maturation is especially important for concentration, sitting quietly, reading, and writing—all skills that are required for formal education. (textbook, p. 182)

14. **b.** Many male-female differences may be caused by varying amounts of practice on certain skills, rather than by intrinsic body or brain differences. (textbook, p. 193)

15. **d.** This is the principle of cephalo-caudal (head-down) development. (textbook, p. 182)

References

Erikson, Erik. (1977). *Toys and reasons: Stages in the ritualization of experience.* New York: Norton.

Smith, Peter K., (1984). *Play in animals and humans.* Oxford: Basil Blackwell.

These two books discuss play as an essential part of the life of children. Smith's book is a collection of fourteen articles on topics such as imaginary playmates and the costs and benefits of play.

The Play Years: Cognitive Development

ORIENTATION

Each season of life has its particular perspective on the world. At no age is this more apparent than during the play years. Young children think and speak very differently from older children and adults. The most significant cognitive advance during early childhood is the attainment of **symbolic thought,** which opens up for the child a whole new world of imitation and pretend play. At the same time, rapid growth occurs in the use of language, including its grammar, vocabulary, and practical use in conversation. Preschoolers' thought, however, is preoperational thought: the child's thinking is not very logical, and is based on an **egocentric** view of the world.

In addition to exploring these cognitive and linguistic changes that occur in children between the ages of 2 and 6, Chapter 9 of the textbook takes a closer look at preschool education. In the past, most children remained home until about 6, but today most begin their schooling at an earlier age. How are today's children affected by their preschool experiences?

Audio program 9, "Then Sentences," picks up where program 6, "First Words," left off in describing the path children follow in acquiring language. With the expert commentary of psycholinguist Jill de Villiers, the audio program examines how children come to produce their first primitive sentences and then move on to produce more complex, grammatically correct speech.

The journey from cries to words to articulate speech is an intricate one unique to human beings. The similar developmental course of children the world over points to the importance of the biological clock in language development. The grammatical errors, abbreviations, and overextensions of rules in children's speech suggest that humans have a natural propensity to acquire language and that children master its complicated rules by actively experimenting with them, rather than merely by imitating the speech that they hear.

As the program opens, we hear audio snapshots of one child at three different ages. The snapshots reveal the remarkable transition from words to sentences.

LESSON GOALS

By the end of this lesson you should be prepared to:

1. Describe how preschool children think.

2. Outline the main accomplishments and limitations of language development during the play years.

3. Discuss why the play years are a prime period for learning, and identify the kinds of experiences that best foster cognitive development during early childhood.

Audio Assignment

Listen to the audio tape that accompanies Lesson 9: "Then Sentences."

Write answers to the following questions. You may replay portions of the program if you need to refresh your memory. Answer guidelines may be found in the Lesson Guidelines section at the end of this chapter.

1. Outline the course of language development during the year after the child's production of his or her first word.

2. What evidence is there of a common timetable for language development determined by the biological clock?

3. Describe the ways in which young children and adults speak to each other and how this interaction promotes the child's acquisition of the rules of language.

4. Compare and contrast the state of readiness of the auditory system, nervous system, and vocal tract for language acquisition during early childhood.

5. What evidence is presented in the program that humans have a natural propensity for learning sign language?

Textbook Assignment

Read Chapter 9: "The Play Years: Cognitive Development," pages 197–217 in *The Developing Person Through the Life Span, 2/e.*

Write your answers to the following questions. Refer back to the textbook, if necessary. Answer guidelines may be found in the Lesson Guidelines section at the end of this chapter.

1. Explain how the cognitive potential of young children is expanded by the ability to use symbolic thinking.

2. Identify and describe several characteristics of preoperational thought.

3. What recent evidence has led to a revision of Piaget's description of cognitive development in the play years?

4. Give several examples of egocentrism in language use and other characteristics of preoperational cognition that make communication more difficult for children.

5. Cite several factors that result in individual differences in the development of language skills.

Testing Yourself

After you have completed the audio and text review questions, see how well you do on the following quiz. Correct answers, with text and audio references, may be found at the end of this chapter.

1. Concerning the acquisition of language during early childhood, which of the following is true?
   a. The first true sentences do not occur until about age 3.
   b. Children's two-word sentences show a lack of knowledge of grammatical rules.

    c. Grammatical rules are evident even in the first two-word sentences of children.

    d. In acquiring language, children merely imitate the speech they hear.

2. Children's abbreviated "telegrams" communicated in their two-word sentences show that:

    a. young children do not yet possess a knowledge of grammar.

    b. they are actively experimenting with the rules of grammar.

    c. children are imitating the sloppy grammar of adults they listen to.

    d. adults tend to rush the acquisition of language before children are truly ready.

3. In terms of producing readiness for hearing and producing sentences, which is the correct order in which the three structures indicated mature?

    a. Auditory system; nervous system; vocal tract

    b. Nervous system; auditory system; vocal tract

    c. Vocal tract; auditory system; nervous system

    d. Auditory system; vocal tract; nervous system

4. Children usually produce their first two-word sentences at about age:

    a. one year.

    b. one-and-a-half or two.

    c. two-and-a-half or three.

    d. three-and-a-half or four.

5. Concerning the acquisition of the rules of language, which of the following is true?

    a. Rule learning is quicker with sign language than with spoken language.

    b. Children learn the rules of language without being explicitly taught.

    c. Two-word sentences appear at the same time in all the world's cultures.

    d. All of the above are true.

6. The most significant cognitive development during the play years is the beginning of:

    a. symbolic thought.

    b. egocentrism.

    c. reversibility.

    d. formal operations.

7. According to Piaget, egocentrism refers to:

    a. the central role of the ego in language acquisition.

    b. the child's inability to realize that things continue to exist even when out of sight.

    c. the way a child's ideas about the world are limited by his or her own point of view.

    d. young children's preoccupation with obtaining immediate gratification of their needs.

8. A child who is unable to grasp the principle of conservation would be unable to:

    a. see things from another person's viewpoint.

    b. retain earlier schemas when confronted by new experiences.

c. reverse logical operations.

d. recognize that the quantity of a substance remains the same despite changes in its shape.

9. To understand that $1 + 2 = 3$ is the same as $3 - 2 = 1$, a child will have had to master the concept of:

a. animism.

b. reversibility.

c. conservation.

d. egocentrism.

10. Preschoolers who apply the rules of grammar when they should not are demonstrating:

a. egocentrism.

b. pragmatic thought.

c. animism.

d. overregularization.

11. One characteristic of preoperational thought is:

a. being able to categorize objects.

b. the ability to count in multiples of 5.

c. being unable to perform logical operations.

d. difficulty adjusting to changes in routine.

12. A preschooler who focuses his or her attention on only one feature of a situation is demonstrating a characteristic of preoperational thought called:

a. centration.

b. pragmatic thinking.

c. reversibility.

d. centrism.

13. The vocabulary of preschool children consists mostly of:

a. metaphors and idioms.

b. verbs and concrete nouns.

c. made up words.

d. abstract nouns and adjectives.

14. Long-term studies of Headstart graduates show that:

a. they achieve more, but have more behavioral problems.

b. they have significantly higher IQ scores.

c. they have higher aspirations and attain higher scores on achievement tests.

d. the benefits are only short-lived.

15. Critics of Piaget's "three mountain experiment" argue that:

a. the inconsistency in results from child to child negates its validity.

b. the task does not consider cultural influences on cognitive development.

c. most children today cannot solve the problem until a later age than Piaget believed.

d. the task is too complex to be a valid test of the child's ability to take another's perspective.

LESSON 9 STUDENT EXERCISE: PRESCHOOL LITERATURE

You can learn much about the language and thought processes of young children by examining the literature written for them. Obtain a "classic" or well-loved storybook written for children from 3 to 5 years old. Some possibilities include the *I Can Read* books by Arnold Lobel, the *Amelia Bedelia* books by Peggy Parish, the *Mr.* books by Roger Hargreaves, and books for younger children by such well-known authors as Maurice Sendak, Charlotte Zolotow, and Dr. Seuss.

Examine the book carefully. If possible, read it aloud to a child or someone else. Then complete the **Exercise Response Sheet** on the following page and return the sheet to your instructor.

LESSON GUIDELINES

Audio Question Guidelines

1. The first word is usually said around the time of the first birthday, the first two-word sentence between one-and-a-half and two years.

   Two-word utterances are not always sentences, however. The words must appear in isolation, and in combination with one another to produce different meanings in order to qualify as true sentences.

   Early sentences are in effect "telegrams," in that they omit articles, conjunctions, prepositions, and other parts of speech that are not essential to meaning.

   During the next year the child actively experiments with the grammatical rules of language. Vocabulary increases rapidly and the two-word sentences soon become three- and four-word sentences.

   The slow-to-develop vocal tract of humans results in the same accommodations to language in children throughout the world. These include reducing consonant clusters to a single consonant and a preference for certain kinds of pronunciations.

2. The evidence for a common biological timetable in the maturation of language includes the fact that the stage of two-word sentences comes at about the same time in every culture.

   The fact that deaf and hearing children babble at about the same age also suggests a maturational basis for language development.

   The slow maturation of the vocal tract of humans results in the same accommodations in language pronunciation in children of similar age throughout the world.

3. The two- and three-word "telegrams" of children during these years are abbreviations of adult speech in which articles, conjunctions, prepositions, and other unessential words are dropped.

   At the same time that children abbreviate their speech, adults often restate their ideas and expand them into complete, grammatically correct utterances. This process implicitly calls the child's attention to the rules of language and indicates that children are experimenting with language, rather than merely imitating.

   Another illustration of this creative and experimental process comes from the **overextensions** of grammatical rules that are typical of children during this stage.

4. The auditory system is mature and ready for hearing at birth.

   Between one-and-a-half and two years of age the brain and nervous system reach a point of maturation that permits the combination of words into primitive sentences.

   The vocal tract lags behind the auditory system and nervous system in maturing. Until it matures, children are unable to articulate complex sounds.

5. The evidence that humans have a natural propensity for acquiring sign language includes the fact that although deaf and hearing children babble at about the same age, deaf children produce their first signs sooner than hearing children produce their first words. This developmental advantage is maintained when it comes to forming two-word or two-sign combinations.

Textbook Question Guidelines

1. The most significant cognitive advance during early childhood is the emergence of **symbolic thought,** which enables the child to turn a word or object into a symbol of something else.

   Both language and imagination become tools of thought, making the typical 2-year-old much more verbal and creative than the 1-year-old.

   Symbolic thought makes pretend play possible.

   Symbolic thought develops sequentially, becoming increasingly complex as the child becomes able to coordinate mentally an increasing number of schemas.

2. Although preschool children can think symbolically, they cannot perform logical **operations,** or "schemes of connected relational meaning." One such operation is reversibility: preschoolers do not understand that reversing a process will restore the original conditions.

   Preoperational children also fail to answer correctly problems involving **conservation,** indicating they do not yet have the idea that the amount of a substance is unaffected by changes in its shape or placement.

   Centration refers to the tendency of preschoolers to focus, or center, on one aspect of a problem, making it difficult for them to understand problems involving cause and effect.

   Egocentrism limits a child's ideas about the world to his or her own point of view.

   Animism refers to the belief of preschool children that everything in the world is alive, just as they are.

3. Recent experiments have shown that under certain circumstances preschoolers may show some of the cognitive capacities—including perspective-taking and conservation—that Piaget's tests suggest are not acquired until the school years.

   In one experiment, children were asked to hide a little figure behind a series of walls so that other figures could not see it. Children as young as three-and-a-half to five years of age were able to hide the figure correctly.

   Simplifying tasks, encouraging the child's active participation, and making the experiment a game rather than a test also seem to help children grasp the concept of conservation at an earlier age.

4. Piaget found that about half of the statements made by preschoolers were egocentric and made no attempt to consider another's viewpoint.

   Monologue speech occurs when children talk to themselves, or others, without waiting for a response. Preschoolers often engage in a **collective monologue** in which they interact conversationally but do not listen or respond to what the others say.

   Preschoolers tend to overestimate the clarity of their communications and understanding, often jumping to erroneous conclusions based on poorly comprehended material.

   Because preoperational thought is concrete, vocabulary is generally limited to concrete nouns and adjectives.

   As an example of "verbal centering," preschoolers tend to take everything literally.

   Due to their difficulty in performing certain operations, preoperational children have difficulty with words that express comparisons or relationships.

5. By the time children enter kindergarten, individual differences in the use of language are apparent.

   Girls, middle-class children, first-borns, and only children tend to be more proficient in language production than boys, lower-class children, and later-borns, respectively.

   Individual variation within groups is, however, greater than variation between groups.

   Middle-class parents tend to give children more elaborate explanations and comments than do lower-class parents.

   Parents tend to talk more to first-borns and single-borns than to later-borns or twins.

   In cultures that do not encourage children's use of language, children do poorly on tests of language development.

Answers to Testing Yourself

1. **c.** Even the two- and three-word "telegrams" of preschoolers show evidence of a rudimentary understanding of grammar. (audio program)

2. **b.** Children master the rules of grammar by experimenting and testing hypotheses, often overextending rules and making errors that cannot be attributed to imitation. (audio program)

3. **a.** The slow-to-mature vocal tract forces accommodations in pronunciation that are similar the world over. (audio program)

4. **b.** Although there is wide variation in normal development, the first primitive sentences usually occur between one-and-a-half and two years of age. (audio program)

5. **d.** All of these statements are true, indicating a common timetable in the maturation of language and the natural propensity children have for acquiring language, including sign language. (audio program)

6. **a.** Symbolic thought results in the rapid expansion of the preschooler's ability to understand, imagine, and communicate. (textbook, p. 198)

7. **c.** The egocentric child does not take into account the fact that other people may have thoughts and feelings different from the ones he or she is having. (textbook, p. 202)

8. **d.** The lack of comprehension of conservation is a characteristic of preoperational thought. (textbook, p. 201)

9. **b.** Reversibility is the concept that reversing a process will restore the original conditions from which the process began. (textbook, p. 199)

10. **d.** Although technically an error, overregularization is actually a sign that children are applying rules of grammar. (textbook, pp. 208–209)

11. **c.** In other words, preschoolers cannot regularly apply relational rules, or operations, such as "if this, then that." (textbook, p. 199)

12. **a.** This tendency to think about one idea at a time is one of the most notable characteristics of preoperational thought. (textbook, p. 200)

13. **b.** This limitation in the language of the preschool child results from the fact that preoperational thinking is concrete, emphasizing appearances and specifics rather than abstractions and hypotheticals. (textbook, p. 211)

14. **c.** Headstart graduates score higher on achievement tests, have better report cards, and higher aspirations than children from the same backgrounds and neighborhoods who did not participate in Headstart. (textbook, p. 215)

15. **d.** Simpler experiments than the three-mountains task have shown that even preschool children can begin to understand the viewpoint of another person. (textbook, pp. 203–204)

References

de Villiers, Jill G. (1980). The process of rule learning in children: A new look. In K. E. Nelson (Ed.), *Children's language* (Vol. 2). New York: Gardner Press.

Professor de Villiers, who is heard on the audio program, discusses the process by which children master the grammatical rules of language.

The Play Years: Psychosocial Development

AUDIO PROGRAM: **Because I Wear Dresses**

ORIENTATION

As we learned in Lessons 8 and 9, the physical and cognitive development that occurs between the ages of 2 and 6 is extensive. Body proportions begin to resemble those of adults; language develops rapidly; and there is an increasing capacity to use mental representation and symbols. Lesson 10 concludes the unit on the play years by exploring ways in which preschool children relate to others in their ever-widening social environment.

During the preschool years a child's self-confidence, social skills, and social roles become more fully developed. This growth coincides with the child's increased capacity for communication, imagination, and understanding of his or her social context. Chapter 10 of *The Developing Person Through the Life Span, 2/e,* examines these aspects of development and discusses several important influences during this season, including parental style, social play, and television viewing. The chapter also examines sex-role development and the controversial theory of **androgyny.** Rather than adhering to traditional sex-role stereotypes, androgynous men and women share many of the same personality characteristics.

Audio program 10, "Because I Wear Dresses," focuses on how children develop **gender identity** as boys or girls. Through the expert commentary of psychologists Michael Stevenson and Jacquelynne Eccles, we discover that by the time children begin elementary school they have developed a strong sense of their gender. During these years children segregate themselves according to sex and become quite stereotyped in their thinking and behavior regarding gender.

Are the psychological differences between the sexes a result of our biology or are they something we learn? One way of looking at this question is to examine how sex differences change across the life span. According to psychologist David Gutmann, masculinity and femininity mean different things at different ages. Although sex differences in such characteristics as aggressiveness may have a biological basis, the difference may not be the same for the entire life cycle. Researchers are finding that men and women become more alike in their actions and attitudes as they get older.

Social guidelines for the sexes may be blurred in old age, but if you ask children about gender differences, you are likely to hear the kind of answers that open this program.

LESSON GOALS

By the end of this lesson you should be prepared to:

1. Discuss the development of the self during the preschool years.

2. Discuss the importance of play in the lives of preschoolers.

3. Discuss how various modes of family interaction affect children's development.

4. Describe the transformation of gender identity and sex-typed behavior over the life span.

Audio Assignment

Listen to the audio tape that accompanies Lesson 10: "Because I Wear Dresses."

Write answers to the following questions. You may replay portions of the program if you need to refresh your memory. Answer guidelines may be found in the Lesson Guidelines section at the end of this chapter.

1. Describe the development of gender identity in preschoolers.

2. Explain some of the gender differences that are usually apparent in the first ten years of life.

3. Explain how gender differences in aggression change over the life span.

Textbook Assignment

Read Chapter 10, "The Play Years: Psychosocial Development," pages 219–244, in *The Developing Person Through the Life Span, 2/e.*

Write answers to the following questions. You may refer back to the textbook, if necessary. Answer guidelines may be found in the Lesson Guidelines section at the end of this chapter.

1. Describe the points of agreement among psychoanalytic, cognitive, and learning theories regarding psychosocial development in early childhood.

2. Identify and differentiate among five kinds of play.

3. Identify and describe three styles of parenting and their consequences on children's development.

4. Describe some of the effects of punishment and suggest alternative actions that parents can take in disciplining their children.

5. Outline the benefits and possible harmful effects of television viewing on young children.

6. Describe possible problems that may occur as a result of childhood aggression, fantasy, and fear.

7. Identify two serious psychological disturbances of childhood and outline the best means of alleviating them.

8. Compare and contrast the three theoretical perspectives on sex-role development during the play years.

Testing Yourself

After you have completed the audio and text review questions, see how well you do on the following quiz. Correct answers, with text and audio references, may be found at the end of this chapter.

1. Most children are able to use gender labels accurately by age:
   a. 1.
   b. 3.
   c. 5.
   d. 7.

2. Concerning gender roles during later life, David Gutmann believes that:
   a. gender roles are the same as earlier in life.
   b. during middle and late adulthood, each sex moves toward a middle ground between the traditional gender roles.
   c. once the demands of parenting are removed, traditional gender roles are reestablished.
   d. gender roles are unrelated to cultural experiences.

3. The greater aggressiveness of boys compared to girls is:
   a. due to boys having a higher natural level of testosterone.
   b. found in virtually all known cultures.
   c. maintained throughout the life cycle.
   d. such that a and b are true.

4. Concerning children's concept of gender, which of the following statements is true?
   a. Until the age of 5 or so, children think that boys and girls can change gender as they get older.
   b. Children as young as 18 months have a clear understanding of the anatomical differences between girls and boys.
   c. Children are inaccurate in labeling others' gender until about age 5.
   d. All of the above are true.

5. Which of the following most accurately summarizes the audio program's explanation of the psychological differences between the sexes?

    a. Most sex differences are biologically determined.
    b. Most sex differences are learned.
    c. Most sex differences are jointly determined by learning and biology.
    d. None of the above are true.

6. According to Erik Erikson, the psychosocial crisis of the years between 3 and 6 is one of:

    a. basic trust versus mistrust.
    b. initiative versus guilt.
    c. industry versus inferiority.
    d. identity versus confusion.

7. Two children who are playing with similar toys in similar ways, but are not interacting, are engaged in:

    a. solitary play.
    b. onlooker play.
    c. associative play.
    d. parallel play.

8. Which style of parenting tends to produce children who are the most self-reliant, self-controlled, and content?

    a. Authoritarian
    b. Permissive
    c. Associative
    d. Authoritative

9. People who score high on measures of androgyny:

    a. follow traditional male and female roles.
    b. are more child-centered than those who do not.
    c. also tend to be very self-centered.
    d. are more flexible in their sex roles.

10. In Freud's theory, the Electra complex causes young girls to:

    a. resent their fathers.
    b. resent their mothers.
    c. resent their fathers and mothers.
    d. identify with their fathers rather than mothers.

11. Between the ages of 3 and 7, children are in which of Freud's psychosexual stages?

    a. Oral
    b. Anal
    c. Phallic
    d. Genital

12. Severely disturbed children who are so self-involved that they prefer self-stimulation to stimulation from others are diagnosed as having:

    a. elective mutism.
    b. echolalia.
    c. autism.
    d. childhood schizophrenia.

13. The technique for treating phobias in which the phobic individual slowly becomes accustomed to the feared object or experience is called:

    a. shaping.
    b. modeling.
    c. progressive extinction.
    d. gradual desensitization.

14. The sophisticated form of social play that helps preschoolers try out social roles, express fears, and learn cooperation is:

    a. onlooker play.
    b. cooperative play.
    c. dramatic play.
    d. associative play.

15. Critics of children's television believe that extensive television viewing may cause children to be:

    a. less social and less verbal.
    b. more social but less verbal.
    c. less social but more creative.
    d. more social but less creative.

LESSON 10 EXERCISE: GENDER ROLE DEVELOPMENT

During the play years, children acquire not only their gender identities but also many masculine or feminine behaviors and attitudes. These behaviors and attitudes largely reflect gender roles. A role is a set of social expectations that prescribes how those who occupy the role should act.

To what extent is your own gender identity a reflection of the behaviors modeled by your parents? Have gender roles become less distinct in recent generations? Should parents encourage gender-stereotyped behaviors in their children? These are among the many controversial questions regarding gender roles that researchers today are grappling with.

The exercise for this lesson asks you to reflect on the kinds of gender models your parents were and to ask a friend or relative who is presently in *a different season of life* to do the same. After you and your respondent have completed the Gender Roles Quiz, answer the questions on the **Exercise Response Sheet** and return only the response sheet to your instructor.

Gender Role Quiz: Respondent #1

For each question, check whether the behavior described was more typical of
your mother or father as you were growing up.

	Mother	Father
1. When your family went out, who drove?		X
2. Who filled out the income tax forms?		X
3. Who wrote the "thank you" notes for gifts?	X	
4. Who was more likely to ask, "Where are my socks/stockings?"		X
5. When the car needed to be repaired, who took it to the garage?		X
6. Who did the laundry?	X	
7. Who dusted and vacuumed your house?	X	
8. When you had a fever, who knew where to find the thermometer?		X
9. When the sink needed fixing, who knew where to find the pipe wrench?		X
10. Who knew where the summer clothes were packed away?	X	
11. When you had guests for dinner, who made the drinks?	X	
12. Who watered the house plants?	X	
13. Who mowed the lawn?	X	X
14. When you went on a trip, who packed the suitcases?	X	
15. When you went on a trip, who packed the car?	X	

Source: Adapted from Doyle, J.A. (1985). *Sex and gender: The human experience.*
Dubuque, Iowa: Wm. C. Brown.

Gender Role Quiz: Respondent #2

For each question, check whether the behavior described was more typical of your mother or father as you were growing up.

	Mother	Father
1. When your family went out, who drove?		X
2. Who filled out the income tax forms?		X
3. Who wrote the "thank you" notes for gifts?	X	
4. Who was more likely to ask, "Where are my socks/stockings?"		X
5. When the car needed to be repaired, who took it to the garage?		X
6. Who did the laundry?	X	
7. Who dusted and vacuumed your house?	X	
8. When you had a fever, who knew where to find the thermometer?	X	
9. When the sink needed fixing, who knew where to find the pipe wrench?	X	
10. Who knew where the summer clothes were packed away?	X	
11. When you had guests for dinner, who made the drinks?		X
12. Who watered the house plants?	X	
13. Who mowed the lawn?	X	
14. When you went on a trip, who packed the suitcases?		X
15. When you went on a trip, who packed the car?	X	

Source: Adapted from Doyle, J.A. (1985). *Sex and gender: The human experience.* Dubuque, Iowa: Wm. C. Brown.

LESSON GUIDELINES

Audio Questions

1. It is likely that children begin to recognize the categories of male and female as early as 18 months of age.

    Most 2-year-olds know whether they are boys or girls, but they have not yet mastered gender constancy. They may think that their sex can change when they grow older or wear different clothes.

    Before the age of 5, most children do not understand the anatomical differences between boys and girls. Instead, they are likely to identify the sexes on the basis of hair length, clothing, or whether a person cooks or goes to work.

    By the time children start school they have developed a very strong sense of their own gender identity, and they know that it will remain constant.

2. Although both biology and environment contribute to differences between the sexes, these differences are slight. There is more variation between individuals *of the same sex* than there is between the sexes.

    As infants, boys are more likely than girls to have been born prematurely, to suffer from birth trauma, to show delayed development, and to be subject to colic and nonrhythmic behaviors that make them somewhat harder to deal with.

    Male infants tend to be less easily cuddled, more resistant to being wrapped up, and more active.

    Perhaps as a consequence of their exposure to higher prenatal levels of testosterone, boys are more likely to get into aggressive encounters than girls.

    From very early on, girls may be more sensitive to faces and to language cues than are boys.

    Boys and girls may also develop different play styles as they go through childhood, with a greater emphasis on competition in boys' games and a greater emphasis on cooperation in girls' games.

3. In nearly every known culture, boys play more aggressively than girls. Boys also have higher levels of testosterone, a hormone linked to aggressiveness.

    David Gutmann believes that males evolved into the more aggressive sex because from the standpoint of species survival, men are more expendable than women. Gutmann also believes that women are responsible for instilling in their children a sense of basic trust—a task facilitated by reduced levels of aggression.

    As men and women get beyond what Gutmann calls the "chronic emergency of parenting," changes in their dispositions become evident. In men, there is an ebbing away of aggressiveness and a flowing in of affiliative and nurturant qualities.

In women the reverse occurs. Freed of the responsibility for their children's emotional security, women's natural aggressiveness begins to surface. The net result of these changes is that the two sexes become more alike in later life.

Text Questions

1. Psychoanalytic, cognitive, and learning theories recognize the emerging concept of self in the preschool child.

    In Erikson's theory, the crisis of the play years is **initiative versus guilt.** The child is turning away from an exclusive attachment to parents and moving toward membership in the larger culture.

    Cognitive theory emphasizes how the preschool child's maturing cognitive processes and accumulating social experiences reduce the egocentric sense of self and increase social understanding.

    Learning theorists note that praise and blame become more effective as mechanisms of reinforcement and punishment than they were earlier, because the child has a greater awareness of self and of how others perceive him or her.

2. The different types of play include **solitary play,** in which a child plays alone; **onlooker play,** in which a child watches others play; **parallel play,** in which children interact but do not seem to be playing the same game; and **cooperative play,** in which children truly play together, by taking turns and helping one another.

    Dramatic play, which involves two or more children in a make-believe plot that they create themselves, coincides with the attainment of symbolic thinking.

    Social play provides crucial experiences in early childhood for learning reciprocity, nurturance, and cooperation. Such experiences are difficult to acquire later in the life span.

3. As stressed by the systems approach, the effects of any style of parenting depend on many interacting factors, including the child's age, sex, and temperament; the parents' personalities, history, and economic circumstances; the needs of other family members; and cultural values.

    The three basic patterns of parenting are **authoritarian** parents, who are too strict; **permissive** parents, who are too lax; and **authoritative** parents, who are more democratic in defining and enforcing rules.

    Children of authoritarian parents tend to be distrustful, unhappy, hostile, and are unlikely to be high-achievers.

    Children of permissive parents are the least self-reliant, the least self-controlled, and the most unhappy.

    Children of authoritative parents are the most self-reliant, self-controlled, friendly, cooperative, and high-achieving.

4. Harsh punishment tends to produce children who are temporarily obedient but who become antisocial "problem" children as they grow older.

The children of parents who are overly critical tend to be more withdrawn and anxious as they grow older.

As alternatives to punishment, developmentalists recommend positive reinforcement for good behavior; realistic expectations of children based on their physical, cognitive, and psychological abilities; discussion of rules; and setting a good example.

When punishment is used, it should be consistent and immediate.

If the child breaks rules frequently, parents should consider that the problem may lie in some aspect of the home situation rather than in the child.

5. Critics cite three areas in which television viewing may have a harmful effect on young children: the effect of commercials; the content of programs; and the time that could be better spent on other activities.

Preschool children usually accept commercial messages uncritically due to their preoperational and egocentric thinking. Commercials also tend to reinforce certain social stereotypes.

Many psychologists believe that the violent content of TV programs promotes violence in children, through both modeling of aggressive acts and the child's development of a passive reaction to the facts and consequences of violence.

Research has shown, however, that educational programs such as "Sesame Street" and "Mister Rogers' Neighborhood" do succeed in their teaching efforts.

Some critics maintain that the time spent in watching even "good" television programs robs children of play time, making them less creative, less verbal, less social, less independent, and perhaps even lower achievers than children who watch less TV.

6. As children grow older, the frequency of deliberate physical aggression increases, peaking during the preschool years.

A certain level of aggression is a normal and healthy sign of self-assertion.

A continuing pattern of aggression may be a precursor of delinquency and adult criminality.

Because of their vivid imagination, preschoolers sometimes have nightmares, elaborate daydreams, and imaginary friends or enemies.

The centered, preoperational thought of preschoolers makes it difficult for them to differentiate reality from fantasy.

If children's imaginations lead to the development of **phobias, modeling** and **gradual desensitization** may be useful therapies.

7. **Autism** is the most severe disturbance of early childhood. Autistic children are completely self-involved, perform repetitive behaviors, and prefer self-stimulation to stimulation from others.

Autistic children generally are mute, but sometimes engage in **echolalia,** a pattern of speech in which they echo, word for word, whatever they hear.

Childhood schizophrenia involves a deterioration in thought processes and social skills, unpredictable emotional behavior, and terrifying nightmares.

Recent evidence suggests that both autism and schizophrenia are strongly influenced by congenital factors, including prenatal diseases, postnatal seizures, and early indicators of brain damage.

The most successful treatment methods for these disorders usually combine behavior therapy with individual attention.

8. The three major theories disagree about the origins of sex-role preferences and stereotypes.

Psychoanalytic theory states that between the ages of 3 and 7 children are in the **phallic stage,** during which time the **Oedipus complex** (in boys) and the **Electra complex** as well as **penis envy** (in girls) lead eventually to the child's **identification** with the same-sex parent.

Learning theorists argue that sex roles are instilled because parents and society provide models and reinforcement for appropriate sex-role behavior and punishment for inappropriate behavior, especially in boys.

Cognitive theorists maintain that until children are at least 4 or 5, they do not realize that they are permanently male or female. Once they do, their acquisition of sex roles parallels their passage through the various stages of cognitive development.

The concept of **androgyny** counters the idea that "masculine" and "feminine" characteristics need to be opposites.

This viewpoint emphasizes that both boys and girls can be raised to have many of the same desirable human characteristics.

Because they are more flexible in their sex roles, androgynous people are generally more competent and have higher self-esteem than people whose behavior is determined by traditional sex roles.

Answers to Testing Yourself

1. **b.** Between 2 and 3 years of age, children become very accurate at labeling others as "he" and "she." (audio program)

2. **b.** David Gutmann believes that gender roles become less distinct as we grow older. (audio program)

3. **d.** Boys have a greater natural endowment of testosterone and are more aggressive in virtually all cultures. (audio program)

4. **a.** Before age 5, many children think their gender may change as they get older. (audio program)

5. **c.** Most psychological differences between women and men are jointly determined by learning and biology. (audio program)

6. **b.** According to Erikson, the crisis of the play years is initiative versus guilt. (textbook, p. 220)

7. **d.** Just as parallel lines do not intersect, children engaged in parallel play do not interact. (textbook, p. 223)

8. **d.** Children who grow up in authoritative families that give them both love and limits are most likely to become successful and happy with themselves. (textbook, p. 229)

9. **d.** Androgynous men and women share many of the same personality traits. (textbook, p. 242)

10. **b.** According to Freud, at about age 4 girls have sexual feelings for their father and accompanying hostility toward their mother. (textbook, p. 239)

11. **c.** Freud called the period from 3 to 7 the phallic stage, because he believed that its center of focus is the penis. (textbook, p. 238)

12. **c.** "Auto" means self. Autistic children prefer self-stimulation to stimulation from others. (textbook, p. 236)

13. **d.** In this technique the child gradually becomes accustomed, or desensitized, to the feared object or situation. (textbook, p. 236)

14. **c.** Dramatic play is mutual fantasy play in which children choose roles and cooperate in acting them out. (textbook, pp. 224–227)

15. **a.** Television tends to cut off social communication, which is essential for enhancing social and verbal skills. (textbook, pp. 230–231)

References

Doyle, J.A. (1985). *Sex and gender: The human experience.* Dubuque, Iowa: Wm. C. Brown.

Doyle's book discusses many controversial issues in the development of gender roles and gender identity.

The School Years: Physical Development

AUDIO PROGRAM: Everything is Harder

ORIENTATION

For most boys and girls, the years of middle childhood are a time when physical growth is smooth and uneventful. Body maturation coupled with sufficient practice enables school-age children to master many motor skills. Chapter 11 of *The Developing Person Through the Life Span, 2/e,* outlines physical development during the school years, noting that boys and girls have about the same physical skills during this season. Although malnutrition limits the growth of children in some regions of the world, most of the variations in physical development in North America are due to heredity. Diet does exert its influence, however, by interacting with heredity, activity level, and other factors to promote **obesity**— the most serious growth problem during middle childhood in North America.

The textbook also discusses the needs of children with physical and educational handicaps, concluding with an evaluation of the advantages and disadvantages of educational **mainstreaming** to both normal children and children with special needs.

Audio Program 11, "Everything is Harder," introduces Sean Miller and Jenny Hamburg, each of whom is disabled by **cerebral palsy.** Through their stories, illuminated by the expert commentary of physical rehabilitation specialist Dr. Virginia Nelson, we discover that when physical development does not go as expected, everything is "off-time" and harder for all concerned. For **disabled** children, **handicapped** by the world around them, nothing—from getting around to meeting the ordinary developmental tasks of life—comes smoothly or easily.

LESSON GOALS

By the end of this lesson you should be prepared to:

1. Describe patterns of normal physical development and motor skill acquisition during the school years.

2. Explain the means for preventing and treating childhood obesity.

3. Explain the advantages and disadvantages of mainstreaming disabled children.

4. Discuss the ways in which meeting the developmental tasks of life is more difficult for children with disabilities.

Audio Assignment

Listen to the audio tape that accompanies Lesson 11: "Everything is Harder: Children with Disabilities."

Write answers to the following questions. You may replay portions of the program if you need to refresh your memory. Answer guidelines may be found in the Lesson Guidelines section at the end of this chapter.

1. Identify the causes and characteristics of cerebral palsy.

2. Differentiate physical disabilities from social handicaps and cite several reasons why development is harder for disabled children.

Textbook Assignment

Read Chapter 11, "The School Years: Physical Development," pages 249–259 in *The Developing Person Through the Life Span, 2/e.*

Write answers to the following questions. Refer back to the textbook, if necessary. Answer guidelines may be found in the Lesson Guidelines section at the end of this chapter.

1. Describe normal physical development during middle childhood and account for the usual variations observed among children.

2. Identify several physical and psychological factors related to obesity in children.

3. Describe the development of motor skills (and note limitations) during the school years.

4. Describe two different approaches to educating children with disabilities and note the advantages and disadvantages of each approach.

Testing Yourself

After you have completed the audio and text review questions, see how well you do on the following quiz. Correct answers, with text and audio references, may be found at the end of this chapter.

1. A movement disorder that results from brain injury occurring around the time of birth is called:
   a. epilepsy.
   b. Huntington's disease.
   c. Parkinson's disease.
   d. cerebral palsy.

2. According to experts in the audio program, disabilities are _____ imposed and handicaps are _____ imposed.
   a. physically . . . socially
   b. socially . . . physically
   c. physically . . . physically
   d. socially . . . socially

3. Which of the following was *not* cited in the program as a developmental obstacle faced by disabled children and their parents?
   a. There are few good role models with physical disabilities.
   b. Because many things take more time, difficult choices between activities must sometimes be made.
   c. Parents of disabled children experience many stresses that other parents do not.
   d. Counselors of disabled children are unwilling to let them make their own choices.

4. According to the experts heard in the audio program, the most difficult time of life for a disabled person is likely to be:
   a. early childhood.
   b. early adolescence.
   c. early adulthood.
   d. all seasons of life.

5. Most experts contend that at least _____ percent of American children need to lose weight.
    a. 10
    b. 15
    c. 20
    d. 25

6. Compared to the physical appearance of most preschoolers, school-age children are:
    a. leaner and more muscular.
    b. leaner and less muscular.
    c. heavier and more muscular.
    d. heavier and less muscular.

7. During the school years growth is:
    a. slower than in childhood and adolescence.
    b. slower than in childhood but faster than in adolescence.
    c. faster than in childhood but slower than in adolescence.
    d. faster than in childhood and adolescence.

8. In the United States, the major variable in the differences in height of school-age children is:
    a. nutrition.
    b. heredity.
    c. early activity level.
    d. medical history.

9. According to the textbook, the best way for obese children to lose weight is to:
    a. eat less.
    b. eat less *and* exercise more.
    c. increase physical activity.
    d. join a medical weight-loss program.

10. In relation to weight in later life, childhood weight:
    a. is not an accurate predictor of adolescent or adult weight.
    b. is predictive of adolescent but not adult weight.
    c. is predictive of adult but not adolescent weight.
    d. is predictive of both adolescent and adult weight.

11. Mainstreaming refers to the practice of:
    a. providing separate classrooms for normal and handicapped children.
    b. providing shared classrooms for normal and handicapped children.
    c. teaching normal children about the special needs of handicapped children.
    d. denying that handicapped children are different from normal children.

12. Which of the following is the *least* important factor in the development of motor skills during the school years?
    a. Gender
    b. Body size
    c. Practice
    d. Brain maturation

13. Overfeeding during infancy may:
    a. promote insecure attachment.
    b. increase the number of fat cells in the child's body.
    c. impair motor skill development.
    d. decrease the number of muscle fibers in the child's body.

14. During middle childhood, girls tend to be better than boys at activities requiring _____; boys tend to be better at activities requiring

    _____
    a. endurance . . . strength.
    b. strength . . . endurance.
    c. forearm strength . . . flexibility.
    d. flexibility . . . forearm strength.

15. Which of the following was *not* cited in the textbook as a *common* cause of obesity?
    a. Television-watching
    b. Repeated dieting
    c. Overeating of high-fat foods
    d. Metabolic problems

LESSON 11 EXERCISE: EVERYTHING IS HARDER

A major theme of Lesson 11 is that when physical, cognitive, or psychosocial development does not proceed normally, life becomes more difficult for everyone involved. Parents find that it takes more time and effort to raise a handicapped child and that other people often are hurtful in their comments and behavior toward the child. Disabled children may need extra self-confidence, self-esteem, and persistence to master tasks that are routine for other children. Later, during adolescence, when young people want to be like everyone else, disabled adolescents find that they cannot do the same things, look the same, or keep up with their friends.

Throughout the series, the developmental theory of Erik Erikson has been discussed as an important model for studying life-span changes. As you will recall from Lesson 2, Erikson's theory identifies eight important psychosocial crises in life and, hence, eight stages of development. Each crisis can be resolved either positively, in a growth-promoting way, or negatively, in a way that disrupts healthy development.

The exercise for Lesson 11 asks you to reflect on Erikson's stages of development as they relate to the special problems of people with physical disabilities. How might a disability make it more difficult for the individual to experience a positive outcome for each crisis? Note that although Lesson 11 focuses on development during middle childhood, this exercise requires you to integrate material from earlier seasons of life and to anticipate later developmental issues. Please answer the following questions and return your completed **Exercise Response Sheet** to your instructor.

LESSON GUIDELINES

Audio Question Guidelines

1. Cerebral palsy is a movement disorder that results from a brain injury, usually one that occurs around the time of birth.

   Although the most obvious symptom of cerebral palsy is that parts of the body do not move the way they normally would, there may be other associated disorders. These include seizures, mental retardation, and hearing and vision problems.

   Unlike many disorders, cerebral palsy is nonprogressive—it does not become worse as the person gets older.

2. Physical disabilities are the result of injury or heredity. Handicaps are imposed by society. Social handicaps such as an environmental obstacle that prevents a wheelchair from entering a building or attitudes that are prejudicial are, in many cases, more disabling than physical disabilities.

   There are many reasons why development is harder for children with disabilities. One is the additional stresses the disability places on parents and other family members.

   Another obstacle to development is that there are few good role models for disabled children.

   Children with disabilities may need greater self-confidence, self-esteem, and "stick-to-it-iveness" than normal children simply because most things are more difficult and take longer.

   Social attitudes often become obstacles for disabled individuals, particularly during adolescence when they want to be like everyone else and discover they are not.

Textbook Question Guidelines

1. Compared to other periods of life, physical development in middle childhood is relatively smooth and uneventful.

   Children grow more slowly than they did earlier, or than they will in adolescence, gaining about 5 pounds and 2½ inches per year.

   By age 10 the average child weighs about 70 pounds and measures 54 inches.

   Children become proportionally thinner as they grow taller.

   Muscles, heart, and lungs become stronger.

   Because most North American children are sufficiently well nourished during middle childhood, most of the variation in their size is attributed to heredity. In some regions of the world, most of the variation in size is caused by malnutrition.

2. The most serious size problem during middle childhood in North America is **obesity,** which probably results from the interaction of many factors, including heredity, physical inactivity, and overfeeding in infancy.

Body type, height, bone structure, and patterns of fat distribution are inherited.

High-fat diets promote obesity.

Food-related attitudes may promote obesity. When food is used to console, or as a symbol of affection, it may lead to overconsumption.

During the first two years of life, the number of fat cells may increase in response to overfeeding.

Television-watching is correlated with obesity. While watching television, children burn fewer calories, consume many snacks, and are bombarded with junk food commercials.

Repeated dieting lowers metabolism and, therefore, the number of calories needed to maintain weight.

Physiological problems such as metabolic abnormalities promote obesity in a small percentage of children.

Approximately 10 percent of American children are estimated to be overweight.

Reducing weight is difficult for children, partly because obesity is often fostered by family attitudes and habits that are difficult to change.

The best way to get children to lose weight is to increase their physical activity and change their eating patterns.

Obesity causes many psychological as well as physiological problems, including a lack of friends and low self-esteem.

3. Body maturation coupled with sufficient practice enables school-age children to perform many motor skills.

With few exceptions, boys and girls are about equal in their physical abilities during the school years.

Boys have greater forearm strength and girls have greater overall flexibility.

Motor skill development is related to several factors, including practice, genetic talent, body size, and brain maturation.

Brain maturation is a key factor in **reaction time.** The typical 7-year-old's reaction time is twice that of an adult.

Ideally, the physical activities of elementary-school children should focus on skills that most of them can perform reasonably well, such as kicking and running.

4. **Mainstreaming** refers to the integration of normal and handicapped children in the classroom.

Three arguments for mainstreaming, and against the separate education of special children, are mentioned in the text:

   a. In terms of the development of **social skills,** normal and special children benefit from sharing a classroom.
   b. **Labeling** itself may retard the development of special children.
   c. Segregating special children is a form of **discrimination.**

Despite its many advantages, mainstreaming has not proven to be a simple solution.

Many special children need extensive and expensive support services.

Social skills are not necessarily fostered simply by mixing normal with special children; special children tend to be socially isolated.

Answers to Testing Yourself

1. **d.** Cerebral palsy is a movement disorder resulting from perinatal brain injury. (audio program)

2. **a.** Social handicaps are often more disabling than physical difficulties. (audio program)

3. **d.** Rehabilitation specialist Virginia Nelson encourages her clients to make their own decisions with regard to walking and driving, for example. (audio program)

4. **b.** Because of the particular pressures of adolescence—such as wanting to be like everyone else—this may be the most difficult season for the disabled person. (audio program)

5. **a.** Experts contend that at least 10 percent of children in the United States need to lose weight. (textbook, p. 251)

6. **a.** During the school years children become proportionally thinner and more muscular. (textbook, p. 249)

7. **a.** Children grow more slowly during middle childhood than they did earlier or will in adolescence. (textbook, p. 249)

8. **b.** In some regions of the world, most of the variation is caused by malnutrition. For most children in North America, heredity rather than diet is responsible for most of the variation in physique. (textbook, p. 250)

9. **c.** Since strenuous dieting during childhood can be harmful, the best way to get children to lose weight is to increase their physical activity. (textbook, p. 251)

10. **d.** If children remain obese throughout childhood, they are likely to be obese as adolescents and adults. (textbook, p. 251)

11. **b.** This practice is called mainstreaming because the handicapped children join the main group of normal children. (textbook, p. 257)

12. **a.** Boys and girls are just about equal in physical abilities during the school years. (textbook, p. 254)

13. **b.** During the first two years of life overfeeding may increase the number of fat cells. (textbook, p. 253)

14. **d.** Although boys and girls are just about equal in motor skills during the school years, girls have greater overall flexibility and boys have greater forearm strength. (textbook, p. 254)

15. **d.** Physiological problems account for less than 1 percent of all cases of childhood obesity.

References

Meisel, C. Julius (1986). *Mainstreaming handicapped children: Outcomes, controversies and new directions.* Hillsdale, NJ: Lawrence Erlbaum.

Various chapters in Meisel's book discuss the history of the mainstreaming concept, its advantages, disadvantages, successes, and failures over the past ten years.

LESSON GUIDELINES

Audio Question Guidelines

1. The first word is usually said around the time of the first birthday, the first two-word sentence between one-and-a-half and two years.

   Two-word utterances are not always sentences, however. The words must appear in isolation, and in combination with one another to produce different meanings in order to qualify as true sentences.

   Early sentences are in effect "telegrams," in that they omit articles, conjunctions, prepositions, and other parts of speech that are not essential to meaning.

   During the next year the child actively experiments with the grammatical rules of language. Vocabulary increases rapidly and the two-word sentences soon become three- and four-word sentences.

   The slow-to-develop vocal tract of humans results in the same accommodations to language in children throughout the world. These include reducing consonant clusters to a single consonant and a preference for certain kinds of pronunciations.

2. The evidence for a common biological timetable in the maturation of language includes the fact that the stage of two-word sentences comes at about the same time in every culture.

   The fact that deaf and hearing children babble at about the same age also suggests a maturational basis for language development.

   The slow maturation of the vocal tract of humans results in the same accommodations in language pronunciation in children of similar age throughout the world.

3. The two- and three-word "telegrams" of children during these years are abbreviations of adult speech in which articles, conjunctions, prepositions, and other unessential words are dropped.

   At the same time that children abbreviate their speech, adults often restate their ideas and expand them into complete, grammatically correct utterances. This process implicitly calls the child's attention to the rules of language and indicates that children are experimenting with language, rather than merely imitating.

   Another illustration of this creative and experimental process comes from the **overextensions** of grammatical rules that are typical of children during this stage.

4. The auditory system is mature and ready for hearing at birth.

   Between one-and-a-half and two years of age the brain and nervous system reach a point of maturation that permits the combination of words into primitive sentences.

   The vocal tract lags behind the auditory system and nervous system in maturing. Until it matures, children are unable to articulate complex sounds.

5. The evidence that humans have a natural propensity for acquiring sign language includes the fact that although deaf and hearing children babble at about the same age, deaf children produce their first signs sooner than hearing children produce their first words. This developmental advantage is maintained when it comes to forming two-word or two-sign combinations.

Textbook Question Guidelines

1. The most significant cognitive advance during early childhood is the emergence of **symbolic thought,** which enables the child to turn a word or object into a symbol of something else.

   Both language and imagination become tools of thought, making the typical 2-year-old much more verbal and creative than the 1-year-old.

   Symbolic thought makes pretend play possible.

   Symbolic thought develops sequentially, becoming increasingly complex as the child becomes able to coordinate mentally an increasing number of schemas.

2. Although preschool children can think symbolically, they cannot perform logical **operations,** or "schemes of connected relational meaning." One such operation is reversibility: preschoolers do not understand that reversing a process will restore the original conditions.

   Preoperational children also fail to answer correctly problems involving **conservation,** indicating they do not yet have the idea that the amount of a substance is unaffected by changes in its shape or placement.

   Centration refers to the tendency of preschoolers to focus, or center, on one aspect of a problem, making it difficult for them to understand problems involving cause and effect.

   Egocentrism limits a child's ideas about the world to his or her own point of view.

   Animism refers to the belief of preschool children that everything in the world is alive, just as they are.

3. Recent experiments have shown that under certain circumstances preschoolers may show some of the cognitive capacities—including perspective-taking and conservation—that Piaget's tests suggest are not acquired until the school years.

   In one experiment, children were asked to hide a little figure behind a series of walls so that other figures could not see it. Children as young as three-and-a-half to five years of age were able to hide the figure correctly.

   Simplifying tasks, encouraging the child's active participation, and making the experiment a game rather than a test also seem to help children grasp the concept of conservation at an earlier age.

4. Piaget found that about half of the statements made by preschoolers were egocentric and made no attempt to consider another's viewpoint.

   Monologue speech occurs when children talk to themselves, or others, without waiting for a response. Preschoolers often engage in a **collective monologue** in which they interact conversationally but do not listen or respond to what the others say.

   Preschoolers tend to overestimate the clarity of their communications and understanding, often jumping to erroneous conclusions based on poorly comprehended material.

   Because preoperational thought is concrete, vocabulary is generally limited to concrete nouns and adjectives.

   As an example of "verbal centering," preschoolers tend to take everything literally.

   Due to their difficulty in performing certain operations, preoperational children have difficulty with words that express comparisons or relationships.

5. By the time children enter kindergarten, individual differences in the use of language are apparent.

   Girls, middle-class children, first-borns, and only children tend to be more proficient in language production than boys, lower-class children, and later-borns, respectively.

   Individual variation within groups is, however, greater than variation between groups.

   Middle-class parents tend to give children more elaborate explanations and comments than do lower-class parents.

   Parents tend to talk more to first-borns and single-borns than to later-borns or twins.

   In cultures that do not encourage children's use of language, children do poorly on tests of language development.

Answers to Testing Yourself

1. **c.** Even the two- and three-word "telegrams" of preschoolers show evidence of a rudimentary understanding of grammar. (audio program)

2. **b.** Children master the rules of grammar by experimenting and testing hypotheses, often overextending rules and making errors that cannot be attributed to imitation. (audio program)

3. **a.** The slow-to-mature vocal tract forces accommodations in pronunciation that are similar the world over. (audio program)

4. **b.** Although there is wide variation in normal development, the first primitive sentences usually occur between one-and-a-half and two years of age. (audio program)

5. **d.** All of these statements are true, indicating a common timetable in the maturation of language and the natural propensity children have for acquiring language, including sign language. (audio program)

6. **a.** Symbolic thought results in the rapid expansion of the preschooler's ability to understand, imagine, and communicate. (textbook, p. 198)

7. **c.** The egocentric child does not take into account the fact that other people may have thoughts and feelings different from the ones he or she is having. (textbook, p. 202)

8. **d.** The lack of comprehension of conservation is a characteristic of preoperational thought. (textbook, p. 201)

9. **b.** Reversibility is the concept that reversing a process will restore the original conditions from which the process began. (textbook, p. 199)

10. **d.** Although technically an error, overregularization is actually a sign that children are applying rules of grammar. (textbook, pp. 208–209)

11. **c.** In other words, preschoolers cannot regularly apply relational rules, or operations, such as "if this, then that." (textbook, p. 199)

12. **a.** This tendency to think about one idea at a time is one of the most notable characteristics of preoperational thought. (textbook, p. 200)

13. **b.** This limitation in the language of the preschool child results from the fact that preoperational thinking is concrete, emphasizing appearances and specifics rather than abstractions and hypotheticals. (textbook, p. 211)

14. **c.** Headstart graduates score higher on achievement tests, have better report cards, and higher aspirations than children from the same backgrounds and neighborhoods who did not participate in Headstart. (textbook, p. 215)

15. **d.** Simpler experiments than the three-mountains task have shown that even preschool children can begin to understand the viewpoint of another person. (textbook, pp. 203–204)

References

de Villiers, Jill G. (1980). The process of rule learning in children: A new look. In K. E. Nelson (Ed.), *Children's language* (Vol. 2). New York: Gardner Press.

Professor de Villiers, who is heard on the audio program, discusses the process by which children master the grammatical rules of language.

The Play Years: Psychosocial Development

AUDIO PROGRAM: **Because I Wear Dresses**

ORIENTATION

As we learned in Lessons 8 and 9, the physical and cognitive development that occurs between the ages of 2 and 6 is extensive. Body proportions begin to resemble those of adults; language develops rapidly; and there is an increasing capacity to use mental representation and symbols. Lesson 10 concludes the unit on the play years by exploring ways in which preschool children relate to others in their ever-widening social environment.

During the preschool years a child's self-confidence, social skills, and social roles become more fully developed. This growth coincides with the child's increased capacity for communication, imagination, and understanding of his or her social context. Chapter 10 of *The Developing Person Through the Life Span, 2/e,* examines these aspects of development and discusses several important influences during this season, including parental style, social play, and television viewing. The chapter also examines sex-role development and the controversial theory of **androgyny.** Rather than adhering to traditional sex-role stereotypes, androgynous men and women share many of the same personality characteristics.

Audio program 10, "Because I Wear Dresses," focuses on how children develop **gender identity** as boys or girls. Through the expert commentary of psychologists Michael Stevenson and Jacquelynne Eccles, we discover that by the time children begin elementary school they have developed a strong sense of their gender. During these years children segregate themselves according to sex and become quite stereotyped in their thinking and behavior regarding gender.

Are the psychological differences between the sexes a result of our biology or are they something we learn? One way of looking at this question is to examine how sex differences change across the life span. According to psychologist David Gutmann, masculinity and femininity mean different things at different ages. Although sex differences in such characteristics as aggressiveness may have a biological basis, the difference may not be the same for the entire life cycle. Researchers are finding that men and women become more alike in their actions and attitudes as they get older.

Social guidelines for the sexes may be blurred in old age, but if you ask children about gender differences, you are likely to hear the kind of answers that open this program.

LESSON GOALS

By the end of this lesson you should be prepared to:

1. Discuss the development of the self during the preschool years.

2. Discuss the importance of play in the lives of preschoolers.

3. Discuss how various modes of family interaction affect children's development.

4. Describe the transformation of gender identity and sex-typed behavior over the life span.

Audio Assignment

Listen to the audio tape that accompanies Lesson 10: "Because I Wear Dresses."

Write answers to the following questions. You may replay portions of the program if you need to refresh your memory. Answer guidelines may be found in the Lesson Guidelines section at the end of this chapter.

1. Describe the development of gender identity in preschoolers.

2. Explain some of the gender differences that are usually apparent in the first ten years of life.

3. Explain how gender differences in aggression change over the life span.

Textbook Assignment

Read Chapter 10, "The Play Years: Psychosocial Development," pages 219–244, in *The Developing Person Through the Life Span, 2/e.*

Write answers to the following questions. You may refer back to the textbook, if necessary. Answer guidelines may be found in the Lesson Guidelines section at the end of this chapter.

1. Describe the points of agreement among psychoanalytic, cognitive, and learning theories regarding psychosocial development in early childhood.

2. Identify and differentiate among five kinds of play.

3. Identify and describe three styles of parenting and their consequences on children's development.

4. Describe some of the effects of punishment and suggest alternative actions that parents can take in disciplining their children.

5. Outline the benefits and possible harmful effects of television viewing on young children.

6. Describe possible problems that may occur as a result of childhood aggression, fantasy, and fear.

7. Identify two serious psychological disturbances of childhood and outline the best means of alleviating them.

8. Compare and contrast the three theoretical perspectives on sex-role development during the play years.

Testing Yourself

After you have completed the audio and text review questions, see how well you do on the following quiz. Correct answers, with text and audio references, may be found at the end of this chapter.

1. Most children are able to use gender labels accurately by age:
   a. 1.
   b. 3.
   c. 5.
   d. 7.

2. Concerning gender roles during later life, David Gutmann believes that:
   a. gender roles are the same as earlier in life.
   b. during middle and late adulthood, each sex moves toward a middle ground between the traditional gender roles.
   c. once the demands of parenting are removed, traditional gender roles are reestablished.
   d. gender roles are unrelated to cultural experiences.

3. The greater aggressiveness of boys compared to girls is:
   a. due to boys having a higher natural level of testosterone.
   b. found in virtually all known cultures.
   c. maintained throughout the life cycle.
   d. such that a and b are true.

4. Concerning children's concept of gender, which of the following statements is true?
   a. Until the age of 5 or so, children think that boys and girls can change gender as they get older.
   b. Children as young as 18 months have a clear understanding of the anatomical differences between girls and boys.
   c. Children are inaccurate in labeling others' gender until about age 5.
   d. All of the above are true.

5. Which of the following most accurately summarizes the audio program's explanation of the psychological differences between the sexes?
   a. Most sex differences are biologically determined.
   b. Most sex differences are learned.
   c. Most sex differences are jointly determined by learning and biology.
   d. None of the above are true.

6. According to Erik Erikson, the psychosocial crisis of the years between 3 and 6 is one of:
   a. basic trust versus mistrust.
   b. initiative versus guilt.
   c. industry versus inferiority.
   d. identity versus confusion.

7. Two children who are playing with similar toys in similar ways, but are not interacting, are engaged in:
   a. solitary play.
   b. onlooker play.
   c. associative play.
   d. parallel play.

8. Which style of parenting tends to produce children who are the most self-reliant, self-controlled, and content?
   a. Authoritarian
   b. Permissive
   c. Associative
   d. Authoritative

9. People who score high on measures of androgyny:
   a. follow traditional male and female roles.
   b. are more child-centered than those who do not.
   c. also tend to be very self-centered.
   d. are more flexible in their sex roles.

10. In Freud's theory, the Electra complex causes young girls to:
    a. resent their fathers.
    b. resent their mothers.
    c. resent their fathers and mothers.
    d. identify with their fathers rather than mothers.

11. Between the ages of 3 and 7, children are in which of Freud's psychosexual stages?
    a. Oral
    b. Anal
    c. Phallic
    d. Genital

12. Severely disturbed children who are so self-involved that they prefer self-stimulation to stimulation from others are diagnosed as having:
    a. elective mutism.
    b. echolalia.
    c. autism.
    d. childhood schizophrenia.

13. The technique for treating phobias in which the phobic individual slowly becomes accustomed to the feared object or experience is called:
    a. shaping.
    b. modeling.
    c. progressive extinction.
    d. gradual desensitization.

14. The sophisticated form of social play that helps preschoolers try out social roles, express fears, and learn cooperation is:
    a. onlooker play.
    b. cooperative play.
    c. dramatic play.
    d. associative play.

15. Critics of children's television believe that extensive television viewing may cause children to be:
    a. less social and less verbal.
    b. more social but less verbal.
    c. less social but more creative.
    d. more social but less creative.

LESSON 10 EXERCISE: GENDER ROLE DEVELOPMENT

During the play years, children acquire not only their gender identities but also many masculine or feminine behaviors and attitudes. These behaviors and attitudes largely reflect gender roles. A role is a set of social expectations that prescribes how those who occupy the role should act.

To what extent is your own gender identity a reflection of the behaviors modeled by your parents? Have gender roles become less distinct in recent generations? Should parents encourage gender-stereotyped behaviors in their children? These are among the many controversial questions regarding gender roles that researchers today are grappling with.

The exercise for this lesson asks you to reflect on the kinds of gender models your parents were and to ask a friend or relative who is presently in *a different season of life* to do the same. After you and your respondent have completed the Gender Roles Quiz, answer the questions on the **Exercise Response Sheet** and return only the response sheet to your instructor.

Gender Role Quiz: Respondent #1

For each question, check whether the behavior described was more typical of your mother or father as you were growing up.

	Mother	Father
1. When your family went out, who drove?		
2. Who filled out the income tax forms?		
3. Who wrote the "thank you" notes for gifts?		
4. Who was more likely to ask, "Where are my socks/stockings?"		
5. When the car needed to be repaired, who took it to the garage?		
6. Who did the laundry?		
7. Who dusted and vacuumed your house?		
8. When you had a fever, who knew where to find the thermometer?		
9. When the sink needed fixing, who knew where to find the pipe wrench?		
10. Who knew where the summer clothes were packed away?		
11. When you had guests for dinner, who made the drinks?		
12. Who watered the house plants?		
13. Who mowed the lawn?		
14. When you went on a trip, who packed the suitcases?		
15. When you went on a trip, who packed the car?		

Source: Adapted from Doyle, J.A. (1985). *Sex and gender: The human experience.* Dubuque, Iowa: Wm. C. Brown.

Gender Role Quiz: Respondent #2

For each question, check whether the behavior described was more typical of your mother or father as you were growing up.

	Mother	Father
1. When your family went out, who drove?		
2. Who filled out the income tax forms?		
3. Who wrote the "thank you" notes for gifts?		
4. Who was more likely to ask, "Where are my socks/stockings?"		
5. When the car needed to be repaired, who took it to the garage?		
6. Who did the laundry?		
7. Who dusted and vacuumed your house?		
8. When you had a fever, who knew where to find the thermometer?		
9. When the sink needed fixing, who knew where to find the pipe wrench?		
10. Who knew where the summer clothes were packed away?		
11. When you had guests for dinner, who made the drinks?		
12. Who watered the house plants?		
13. Who mowed the lawn?		
14. When you went on a trip, who packed the suitcases?		
15. When you went on a trip, who packed the car?		

Source: Adapted from Doyle, J.A. (1985). *Sex and gender: The human experience.* Dubuque, Iowa: Wm. C. Brown.

NAME ———————————————— INSTRUCTOR ——————————————

LESSON 10:

Exercise Response Sheet

1. In what seasons of life were your quiz respondents?

2. Is there evidence of gender-stereotyped behaviors in the respondents' answers?

Were items 1, 2, 4, 5, 9, 11, 13, and 15 checked as more typical of fathers? Were items 3, 6, 7, 8, 10, 12, and 14 checked as more typical of mothers? For each respondent, indicate the total number of responses (out of 15) that are *in agreement* with the traditional gender role breakdown in this list.

	Younger Respondent	Older Respondent
Number of items in agreement with traditional gender roles (maximum = 15)		

If there is a difference in responses given by your younger and older respondents, please explain the difference.

3. To what extent do you believe your own gender identity and gender role development were influenced by the behaviors modeled by your parents? In what ways is your own behavior modeled after that of your same-sex parent? In what ways is it different?

4. To what extent is your concept of the ideal person of the opposite sex a reflection of the behaviors modeled by your opposite-sex parent? In what ways is it different?

5. In your estimation, should parents encourage or discourage traditional gender role development in their children? Please explain your reasoning.

LESSON GUIDELINES

Audio Questions

1. It is likely that children begin to recognize the categories of male and female as early as 18 months of age.

   Most 2-year-olds know whether they are boys or girls, but they have not yet mastered gender constancy. They may think that their sex can change when they grow older or wear different clothes.

   Before the age of 5, most children do not understand the anatomical differences between boys and girls. Instead, they are likely to identify the sexes on the basis of hair length, clothing, or whether a person cooks or goes to work.

   By the time children start school they have developed a very strong sense of their own gender identity, and they know that it will remain constant.

2. Although both biology and environment contribute to differences between the sexes, these differences are slight. There is more variation between individuals *of the same sex* than there is between the sexes.

   As infants, boys are more likely than girls to have been born prematurely, to suffer from birth trauma, to show delayed development, and to be subject to colic and nonrhythmic behaviors that make them somewhat harder to deal with.

   Male infants tend to be less easily cuddled, more resistant to being wrapped up, and more active.

   Perhaps as a consequence of their exposure to higher prenatal levels of testosterone, boys are more likely to get into aggressive encounters than girls.

   From very early on, girls may be more sensitive to faces and to language cues than are boys.

   Boys and girls may also develop different play styles as they go through childhood, with a greater emphasis on competition in boys' games and a greater emphasis on cooperation in girls' games.

3. In nearly every known culture, boys play more aggressively than girls. Boys also have higher levels of testosterone, a hormone linked to aggressiveness.

   David Gutmann believes that males evolved into the more aggressive sex because from the standpoint of species survival, men are more expendable than women. Gutmann also believes that women are responsible for instilling in their children a sense of basic trust—a task facilitated by reduced levels of aggression.

   As men and women get beyond what Gutmann calls the "chronic emergency of parenting," changes in their dispositions become evident. In men, there is an ebbing away of aggressiveness and a flowing in of affiliative and nurturant qualities.

In women the reverse occurs. Freed of the responsibility for their children's emotional security, women's natural aggressiveness begins to surface. The net result of these changes is that the two sexes become more alike in later life.

Text Questions

1. Psychoanalytic, cognitive, and learning theories recognize the emerging concept of self in the preschool child.

   In Erikson's theory, the crisis of the play years is **initiative versus guilt.** The child is turning away from an exclusive attachment to parents and moving toward membership in the larger culture.

   Cognitive theory emphasizes how the preschool child's maturing cognitive processes and accumulating social experiences reduce the egocentric sense of self and increase social understanding.

   Learning theorists note that praise and blame become more effective as mechanisms of reinforcement and punishment than they were earlier, because the child has a greater awareness of self and of how others perceive him or her.

2. The different types of play include **solitary play,** in which a child plays alone; **onlooker play,** in which a child watches others play; **parallel play,** in which children interact but do not seem to be playing the same game; and **cooperative play,** in which children truly play together, by taking turns and helping one another.

   Dramatic play, which involves two or more children in a make-believe plot that they create themselves, coincides with the attainment of symbolic thinking.

   Social play provides crucial experiences in early childhood for learning reciprocity, nurturance, and cooperation. Such experiences are difficult to acquire later in the life span.

3. As stressed by the systems approach, the effects of any style of parenting depend on many interacting factors, including the child's age, sex, and temperament; the parents' personalities, history, and economic circumstances; the needs of other family members; and cultural values.

   The three basic patterns of parenting are **authoritarian** parents, who are too strict; **permissive** parents, who are too lax; and **authoritative** parents, who are more democratic in defining and enforcing rules.

   Children of authoritarian parents tend to be distrustful, unhappy, hostile, and are unlikely to be high-achievers.

   Children of permissive parents are the least self-reliant, the least self-controlled, and the most unhappy.

   Children of authoritative parents are the most self-reliant, self-controlled, friendly, cooperative, and high-achieving.

4. Harsh punishment tends to produce children who are temporarily obedient but who become antisocial "problem" children as they grow older.

The children of parents who are overly critical tend to be more withdrawn and anxious as they grow older.

As alternatives to punishment, developmentalists recommend positive reinforcement for good behavior; realistic expectations of children based on their physical, cognitive, and psychological abilities; discussion of rules; and setting a good example.

When punishment is used, it should be consistent and immediate.

If the child breaks rules frequently, parents should consider that the problem may lie in some aspect of the home situation rather than in the child.

5. Critics cite three areas in which television viewing may have a harmful effect on young children: the effect of commercials; the content of programs; and the time that could be better spent on other activities.

   Preschool children usually accept commercial messages uncritically due to their preoperational and egocentric thinking. Commercials also tend to reinforce certain social stereotypes.

   Many psychologists believe that the violent content of TV programs promotes violence in children, through both modeling of aggressive acts and the child's development of a passive reaction to the facts and consequences of violence.

   Research has shown, however, that educational programs such as "Sesame Street" and "Mister Rogers' Neighborhood" do succeed in their teaching efforts.

   Some critics maintain that the time spent in watching even "good" television programs robs children of play time, making them less creative, less verbal, less social, less independent, and perhaps even lower achievers than children who watch less TV.

6. As children grow older, the frequency of deliberate physical aggression increases, peaking during the preschool years.

   A certain level of aggression is a normal and healthy sign of self-assertion.

   A continuing pattern of aggression may be a precursor of delinquency and adult criminality.

   Because of their vivid imagination, preschoolers sometimes have nightmares, elaborate daydreams, and imaginary friends or enemies.

   The centered, preoperational thought of preschoolers makes it difficult for them to differentiate reality from fantasy.

   If children's imaginations lead to the development of **phobias, modeling** and **gradual desensitization** may be useful therapies.

7. **Autism** is the most severe disturbance of early childhood. Autistic children are completely self-involved, perform repetitive behaviors, and prefer self-stimulation to stimulation from others.

   Autistic children generally are mute, but sometimes engage in **echolalia,** a pattern of speech in which they echo, word for word, whatever they hear.

Childhood schizophrenia involves a deterioration in thought processes and social skills, unpredictable emotional behavior, and terrifying nightmares.

Recent evidence suggests that both autism and schizophrenia are strongly influenced by congenital factors, including prenatal diseases, postnatal seizures, and early indicators of brain damage.

The most successful treatment methods for these disorders usually combine behavior therapy with individual attention.

8. The three major theories disagree about the origins of sex-role preferences and stereotypes.

Psychoanalytic theory states that between the ages of 3 and 7 children are in the **phallic stage,** during which time the **Oedipus complex** (in boys) and the **Electra complex** as well as **penis envy** (in girls) lead eventually to the child's **identification** with the same-sex parent.

Learning theorists argue that sex roles are instilled because parents and society provide models and reinforcement for appropriate sex-role behavior and punishment for inappropriate behavior, especially in boys.

Cognitive theorists maintain that until children are at least 4 or 5, they do not realize that they are permanently male or female. Once they do, their acquisition of sex roles parallels their passage through the various stages of cognitive development.

The concept of **androgyny** counters the idea that "masculine" and "feminine" characteristics need to be opposites.

This viewpoint emphasizes that both boys and girls can be raised to have many of the same desirable human characteristics.

Because they are more flexible in their sex roles, androgynous people are generally more competent and have higher self-esteem than people whose behavior is determined by traditional sex roles.

Answers to Testing Yourself

1. **b.** Between 2 and 3 years of age, children become very accurate at labeling others as "he" and "she." (audio program)

2. **b.** David Gutmann believes that gender roles become less distinct as we grow older. (audio program)

3. **d.** Boys have a greater natural endowment of testosterone and are more aggressive in virtually all cultures. (audio program)

4. **a.** Before age 5, many children think their gender may change as they get older. (audio program)

5. **c.** Most psychological differences between women and men are jointly determined by learning and biology. (audio program)

6. **b.** According to Erikson, the crisis of the play years is initiative versus guilt. (textbook, p. 220)

7. **d.** Just as parallel lines do not intersect, children engaged in parallel play do not interact. (textbook, p. 223)

8. **d.** Children who grow up in authoritative families that give them both love and limits are most likely to become successful and happy with themselves. (textbook, p. 229)

9. **d.** Androgynous men and women share many of the same personality traits. (textbook, p. 242)

10. **b.** According to Freud, at about age 4 girls have sexual feelings for their father and accompanying hostility toward their mother. (textbook, p. 239)

11. **c.** Freud called the period from 3 to 7 the phallic stage, because he believed that its center of focus is the penis. (textbook, p. 238)

12. **c.** "Auto" means self. Autistic children prefer self-stimulation to stimulation from others. (textbook, p. 236)

13. **d.** In this technique the child gradually becomes accustomed, or desensitized, to the feared object or situation. (textbook, p. 236)

14. **c.** Dramatic play is mutual fantasy play in which children choose roles and cooperate in acting them out. (textbook, pp. 224–227)

15. **a.** Television tends to cut off social communication, which is essential for enhancing social and verbal skills. (textbook, pp. 230–231)

References

Doyle, J.A. (1985). *Sex and gender: The human experience.* Dubuque, Iowa: Wm. C. Brown.

Doyle's book discusses many controversial issues in the development of gender roles and gender identity.

The School Years: Physical Development

ORIENTATION

For most boys and girls, the years of middle childhood are a time when physical growth is smooth and uneventful. Body maturation coupled with sufficient practice enables school-age children to master many motor skills. Chapter 11 of *The Developing Person Through the Life Span, 2/e,* outlines physical development during the school years, noting that boys and girls have about the same physical skills during this season. Although malnutrition limits the growth of children in some regions of the world, most of the variations in physical development in North America are due to heredity. Diet does exert its influence, however, by interacting with heredity, activity level, and other factors to promote **obesity**— the most serious growth problem during middle childhood in North America.

The textbook also discusses the needs of children with physical and educational handicaps, concluding with an evaluation of the advantages and disadvantages of educational **mainstreaming** to both normal children and children with special needs.

Audio Program 11, "Everything is Harder," introduces Sean Miller and Jenny Hamburg, each of whom is disabled by **cerebral palsy.** Through their stories, illuminated by the expert commentary of physical rehabilitation specialist Dr. Virginia Nelson, we discover that when physical development does not go as expected, everything is "off-time" and harder for all concerned. For **disabled** children, **handicapped** by the world around them, nothing—from getting around to meeting the ordinary developmental tasks of life—comes smoothly or easily.

LESSON GOALS

By the end of this lesson you should be prepared to:

1. Describe patterns of normal physical development and motor skill acquisition during the school years.

2. Explain the means for preventing and treating childhood obesity.

3. Explain the advantages and disadvantages of mainstreaming disabled children.

4. Discuss the ways in which meeting the developmental tasks of life is more difficult for children with disabilities.

Audio Assignment

Listen to the audio tape that accompanies Lesson 11: "Everything is Harder: Children with Disabilities."

Write answers to the following questions. You may replay portions of the program if you need to refresh your memory. Answer guidelines may be found in the Lesson Guidelines section at the end of this chapter.

1. Identify the causes and characteristics of cerebral palsy.

2. Differentiate physical disabilities from social handicaps and cite several reasons why development is harder for disabled children.

Textbook Assignment

Read Chapter 11, "The School Years: Physical Development," pages 249–259 in *The Developing Person Through the Life Span, 2/e.*

Write answers to the following questions. Refer back to the textbook, if necessary. Answer guidelines may be found in the Lesson Guidelines section at the end of this chapter.

1. Describe normal physical development during middle childhood and account for the usual variations observed among children.

2. Identify several physical and psychological factors related to obesity in children.

3. Describe the development of motor skills (and note limitations) during the school years.

4. Describe two different approaches to educating children with disabilities and note the advantages and disadvantages of each approach.

Testing Yourself

After you have completed the audio and text review questions, see how well you do on the following quiz. Correct answers, with text and audio references, may be found at the end of this chapter.

1. A movement disorder that results from brain injury occurring around the time of birth is called:
   a. epilepsy.
   b. Huntington's disease.
   c. Parkinson's disease.
   d. cerebral palsy.

2. According to experts in the audio program, disabilities are _____ imposed and handicaps are _____ imposed.
   a. physically . . . socially
   b. socially . . . physically
   c. physically . . . physically
   d. socially . . . socially

3. Which of the following was *not* cited in the program as a developmental obstacle faced by disabled children and their parents?
   a. There are few good role models with physical disabilities.
   b. Because many things take more time, difficult choices between activities must sometimes be made.
   c. Parents of disabled children experience many stresses that other parents do not.
   d. Counselors of disabled children are unwilling to let them make their own choices.

4. According to the experts heard in the audio program, the most difficult time of life for a disabled person is likely to be:
   a. early childhood.
   b. early adolescence.
   c. early adulthood.
   d. all seasons of life.

5. Most experts contend that at least _____ percent of American children need to lose weight.
   a. 10
   b. 15
   c. 20
   d. 25

6. Compared to the physical appearance of most preschoolers, school-age children are:
   a. leaner and more muscular.
   b. leaner and less muscular.
   c. heavier and more muscular.
   d. heavier and less muscular.

7. During the school years growth is:
   a. slower than in childhood and adolescence.
   b. slower than in childhood but faster than in adolescence.
   c. faster than in childhood but slower than in adolescence.
   d. faster than in childhood and adolescence.

8. In the United States, the major variable in the differences in height of school-age children is:
   a. nutrition.
   b. heredity.
   c. early activity level.
   d. medical history.

9. According to the textbook, the best way for obese children to lose weight is to:
   a. eat less.
   b. eat less *and* exercise more.
   c. increase physical activity.
   d. join a medical weight-loss program.

10. In relation to weight in later life, childhood weight:
    a. is not an accurate predictor of adolescent or adult weight.
    b. is predictive of adolescent but not adult weight.
    c. is predictive of adult but not adolescent weight.
    d. is predictive of both adolescent and adult weight.

11. Mainstreaming refers to the practice of:
    a. providing separate classrooms for normal and handicapped children.
    b. providing shared classrooms for normal and handicapped children.
    c. teaching normal children about the special needs of handicapped children.
    d. denying that handicapped children are different from normal children.

12. Which of the following is the *least* important factor in the development of motor skills during the school years?
    a. Gender
    b. Body size
    c. Practice
    d. Brain maturation

13. Overfeeding during infancy may:
    a. promote insecure attachment.
    b. increase the number of fat cells in the child's body.
    c. impair motor skill development.
    d. decrease the number of muscle fibers in the child's body.

14. During middle childhood, girls tend to be better than boys at activities requiring _____; boys tend to be better at activities requiring
    _____
    a. endurance . . . strength.
    b. strength . . . endurance.
    c. forearm strength . . . flexibility.
    d. flexibility . . . forearm strength.

15. Which of the following was *not* cited in the textbook as a *common* cause of obesity?
    a. Television-watching
    b. Repeated dieting
    c. Overeating of high-fat foods
    d. Metabolic problems

LESSON 11 EXERCISE: EVERYTHING IS HARDER

A major theme of Lesson 11 is that when physical, cognitive, or psychosocial development does not proceed normally, life becomes more difficult for everyone involved. Parents find that it takes more time and effort to raise a handicapped child and that other people often are hurtful in their comments and behavior toward the child. Disabled children may need extra self-confidence, self-esteem, and persistence to master tasks that are routine for other children. Later, during adolescence, when young people want to be like everyone else, disabled adolescents find that they cannot do the same things, look the same, or keep up with their friends.

Throughout the series, the developmental theory of Erik Erikson has been discussed as an important model for studying life-span changes. As you will recall from Lesson 2, Erikson's theory identifies eight important psychosocial crises in life and, hence, eight stages of development. Each crisis can be resolved either positively, in a growth-promoting way, or negatively, in a way that disrupts healthy development.

The exercise for Lesson 11 asks you to reflect on Erikson's stages of development as they relate to the special problems of people with physical disabilities. How might a disability make it more difficult for the individual to experience a positive outcome for each crisis? Note that although Lesson 11 focuses on development during middle childhood, this exercise requires you to integrate material from earlier seasons of life and to anticipate later developmental issues. Please answer the following questions and return your completed **Exercise Response Sheet** to your instructor.

NAME ———————————————   INSTRUCTOR ———————————————

LESSON 11:

Exercise Response Sheet

How might a disability make it more difficult to develop a positive outcome for each of Erik Erikson's eight psychosocial crises? Base your answers on material from the audio program, your own experiences, or the experiences of someone you know.

Trust vs. mistrust:

Autonomy vs. shame and doubt:

Initiative vs. guilt:

Industry vs. inferiority:

Identity vs. role confusion:

Intimacy vs. isolation:

Generativity vs. self-absorption:

Integrity vs. despair:

LESSON GUIDELINES

Audio Question Guidelines

1. Cerebral palsy is a movement disorder that results from a brain injury, usually one that occurs around the time of birth.

   Although the most obvious symptom of cerebral palsy is that parts of the body do not move the way they normally would, there may be other associated disorders. These include seizures, mental retardation, and hearing and vision problems.

   Unlike many disorders, cerebral palsy is nonprogressive—it does not become worse as the person gets older.

2. Physical disabilities are the result of injury or heredity. Handicaps are imposed by society. Social handicaps such as an environmental obstacle that prevents a wheelchair from entering a building or attitudes that are prejudicial are, in many cases, more disabling than physical disabilities.

   There are many reasons why development is harder for children with disabilities. One is the additional stresses the disability places on parents and other family members.

   Another obstacle to development is that there are few good role models for disabled children.

   Children with disabilities may need greater self-confidence, self-esteem, and "stick-to-it-iveness" than normal children simply because most things are more difficult and take longer.

   Social attitudes often become obstacles for disabled individuals, particularly during adolescence when they want to be like everyone else and discover they are not.

Textbook Question Guidelines

1. Compared to other periods of life, physical development in middle childhood is relatively smooth and uneventful.

   Children grow more slowly than they did earlier, or than they will in adolescence, gaining about 5 pounds and 2½ inches per year.

   By age 10 the average child weighs about 70 pounds and measures 54 inches.

   Children become proportionally thinner as they grow taller.

   Muscles, heart, and lungs become stronger.

   Because most North American children are sufficiently well nourished during middle childhood, most of the variation in their size is attributed to heredity. In some regions of the world, most of the variation in size is caused by malnutrition.

2. The most serious size problem during middle childhood in North America is **obesity,** which probably results from the interaction of many factors, including heredity, physical inactivity, and overfeeding in infancy.

    Body type, height, bone structure, and patterns of fat distribution are inherited.

    High-fat diets promote obesity.

    Food-related attitudes may promote obesity. When food is used to console, or as a symbol of affection, it may lead to overconsumption.

    During the first two years of life, the number of fat cells may increase in response to overfeeding.

    Television-watching is correlated with obesity. While watching television, children burn fewer calories, consume many snacks, and are bombarded with junk food commercials.

    Repeated dieting lowers metabolism and, therefore, the number of calories needed to maintain weight.

    Physiological problems such as metabolic abnormalities promote obesity in a small percentage of children.

    Approximately 10 percent of American children are estimated to be overweight.

    Reducing weight is difficult for children, partly because obesity is often fostered by family attitudes and habits that are difficult to change.

    The best way to get children to lose weight is to increase their physical activity and change their eating patterns.

    Obesity causes many psychological as well as physiological problems, including a lack of friends and low self-esteem.

3. Body maturation coupled with sufficient practice enables school-age children to perform many motor skills.

    With few exceptions, boys and girls are about equal in their physical abilities during the school years.

    Boys have greater forearm strength and girls have greater overall flexibility.

    Motor skill development is related to several factors, including practice, genetic talent, body size, and brain maturation.

    Brain maturation is a key factor in **reaction time.** The typical 7-year-old's reaction time is twice that of an adult.

    Ideally, the physical activities of elementary-school children should focus on skills that most of them can perform reasonably well, such as kicking and running.

4. **Mainstreaming** refers to the integration of normal and handicapped children in the classroom.

Three arguments for mainstreaming, and against the separate education of special children, are mentioned in the text:

   a. In terms of the development of **social skills,** normal and special children benefit from sharing a classroom.
   b. **Labeling** itself may retard the development of special children.
   c. Segregating special children is a form of **discrimination.**

Despite its many advantages, mainstreaming has not proven to be a simple solution.

Many special children need extensive and expensive support services.

Social skills are not necessarily fostered simply by mixing normal with special children; special children tend to be socially isolated.

Answers to Testing Yourself

1. **d.** Cerebral palsy is a movement disorder resulting from perinatal brain injury. (audio program)

2. **a.** Social handicaps are often more disabling than physical difficulties. (audio program)

3. **d.** Rehabilitation specialist Virginia Nelson encourages her clients to make their own decisions with regard to walking and driving, for example. (audio program)

4. **b.** Because of the particular pressures of adolescence—such as wanting to be like everyone else—this may be the most difficult season for the disabled person. (audio program)

5. **a.** Experts contend that at least 10 percent of children in the United States need to lose weight. (textbook, p. 251)

6. **a.** During the school years children become proportionally thinner and more muscular. (textbook, p. 249)

7. **a.** Children grow more slowly during middle childhood than they did earlier or will in adolescence. (textbook, p. 249)

8. **b.** In some regions of the world, most of the variation is caused by malnutrition. For most children in North America, heredity rather than diet is responsible for most of the variation in physique. (textbook, p. 250)

9. **c.** Since strenuous dieting during childhood can be harmful, the best way to get children to lose weight is to increase their physical activity. (textbook, p. 251)

10. **d.** If children remain obese throughout childhood, they are likely to be obese as adolescents and adults. (textbook, p. 251)

11. **b.** This practice is called mainstreaming because the handicapped children join the main group of normal children. (textbook, p. 257)

12. **a.** Boys and girls are just about equal in physical abilities during the school years. (textbook, p. 254)

13. **b.** During the first two years of life overfeeding may increase the number of fat cells. (textbook, p. 253)

14. **d.** Although boys and girls are just about equal in motor skills during the school years, girls have greater overall flexibility and boys have greater forearm strength. (textbook, p. 254)

15. **d.** Physiological problems account for less than 1 percent of all cases of childhood obesity.

References

Meisel, C. Julius (1986). *Mainstreaming handicapped children: Outcomes, controversies and new directions.* Hillsdale, NJ: Lawrence Erlbaum.

Various chapters in Meisel's book discuss the history of the mainstreaming concept, its advantages, disadvantages, successes, and failures over the past ten years.

The School Years: Cognitive Development

AUDIO PROGRAM: **Piaget and the Age of Reason**

ORIENTATION

Cognitive development between the ages of 6 and 11 is impressive, as attested to by children's reasoning strategies, mastery of school-related skills, and use of language. Lesson 12 explores these changes and their significance to the developing person.

Chapter 12 of the textbook begins with a description of Piaget's theory of cognitive development during middle childhood, beginning with the **5-to-7 shift** from **preoperational** to **concrete operational** thinking. When this transition is complete, children are much better able to understand logical principles, as long as they are applied to concrete examples.

The text also notes that the **information-processing** view of cognitive development places more emphasis than Piaget does on the ways in which children process their experiences. During the school years, for example, children become better able to receive, store, and organize information, in part because of their improved memory.

Linguistic development during these years is also extensive, with children showing improvement in vocabulary, grammar, and pragmatic use of language. This is clearly indicated by their newly found delight with words and their growing sophistication in telling jokes.

Audio program 12, "Piaget and the Age of Reason," focuses on a description and critique of Piaget's stages of **preoperational** and **concrete operational** thought. Piaget's famous **conservation** experiments are illustrated with children of several ages, and expert commentary is provided by psychologist David Elkind. Several landmarks of the transition from preoperational to concrete operational thought are illustrated, including the disappearance of egocentric thinking, and the emergence of abilities to classify, deal with rules, consider two dimensions at once, and take another's perspective. Professor Elkind also discusses the contemporary concept of the **competent child** and why the efforts of many modern parents to accelerate cognitive development in their children may be futile.

As the program opens we hear the voices of two children—one who has not yet attained what philosophers once called "The Age of Reason," and one who has.

LESSON GOALS

By the end of this lesson you should be prepared to:

1. Describe how children's cognitive and language abilities change between the ages of 6 and 12.

2. Outline the information-processing view of cognitive development during the school years.

3. Discuss the logical structures of concrete operational thought, according to Piaget.

4. Discuss the possible causes and treatment of specific learning disabilities and the attention deficit disorder.

Audio Assignment

Listen to the audio tape that accompanies Lesson 12: "Piaget and the Age of Reason."

Write answers to the following questions. You may replay portions of the program if you need to refresh your memory. Answer guidelines may be found in the Lesson Guidelines section at the end of this chapter.

1. Identify and describe the major characteristics of preoperational thinking.

2. Explain the conservation-of-liquid task and identify the cognitive abilities that enable children to succeed in this task.

3. Identify the cognitive gains that come with concrete operational thought. In what ways is thinking still limited among children in this stage?

4. Discuss some of the psychological and social effects of preoperational and concrete operational thought on children.

5. Discuss how some of Piaget's ideas have been modified recently, and how the social revolutions of the past 25 years have changed parents' expectations of children.

Textbook Assignment

Read Chapter 12: "The School Years: Cognitive Development," pages 261–289 in *The Developing Person Through the Life Span, 2/e.*

Write your answers to the following questions. Refer back to the textbook, if necessary. Answer guidelines may be found in the Lesson Guidelines section at the end of this chapter.

1. Describe the changes that occur in children's mental abilities between the ages of 5 and 7.

2. Cite several of Piaget's ideas concerning school-age children's thought that remain influential, and several that have been revised by contemporary psychologists.

3. Outline and describe the stages of the information-processing theory.

4. Describe language development during the school years.

5. Describe four specific learning difficulties and discuss their possible causes and treatment.

Testing Yourself

After you have completed the audio and text review questions, see how well you do on the following quiz. Correct answers, with text and audio references, may be found at the end of this chapter.

1. The kind of thinking that does not allow one to see the world from another's point of view is called:
   a. centration.
   b. egocentrism.
   c. concrete operations.
   d. preoperational thought.

2. Although Piaget believed that children could not take another's point of view before the age of _____, some contemporary researchers believe that children as young as age _____ can grasp this concept.
   a. 4 . . . 2
   b. 9 . . . 5
   c. 6 . . . 4
   d. 7 . . . 3

3. Preoperational children lack the concept of conservation because they fail to realize that:
   a. a change in one dimension of an object brings about a change in another dimension too.
   b. matter can neither be created nor destroyed.
   c. rules are not immutable and can be modified.
   d. most conservation experiments distract the child's attention from critical aspects of the problem.

4. The principle that properties such as volume, number, and area remain the same despite changes in the appearance of objects is called:
   a. constancy.
   b. reversibility.
   c. conservation.
   d. reciprocity.

5. Current research on cognitive development indicates that:
   a. Piaget's theory is applicable only to upper-class children.
   b. Piaget overlooked the importance of social development on cognition.
   c. certain cognitive abilities may be acquired at an earlier age than Piaget believed.
   d. Piaget may have overestimated the competence of young children.

6. According to Piaget, a 9-year-old can reason only about concrete things. "Concrete" means:
   a. logical and mathematical.
   b. abstract or vague.
   c. classifiable or listed.
   d. tangible or specific.

7. Of the following tasks, the most difficult for a 6-year-old would probably be:
   a. counting from 1 to 10.
   b. sorting pencils by length.
   c. using polite speech.
   d. telling time from a digital clock.

8. Psychologists who explore the abilities of children to receive, store, and organize information are examining cognitive development from a view based on:
   a. information processing.
   b. Piaget's theory of development.
   c. the influence of the biological clock.
   d. learning theory.

9. Because Karen understands that her cousin is also her grandmother's grandchild, she has mastered the concept of:
   a. reciprocity.
   b. class inclusion.
   c. seriation.
   d. conservation.

10. The ability to use rehearsal, chunking, and other techniques to keep things in mind is called:
    a. metamemory.
    b. sensory register.
    c. selective attention.
    d. pragmatics.

11. Characteristics of Black English include the fact that it:
    a. is not a true language with easily applied rules.
    b. is simply poor standard English and should be discouraged.
    c. presents alternative grammatical forms, some of which can be traced to African linguistic patterns.
    d. is just as difficult to master as any second language.

12. Most children with learning disabilities:
    a. have measurably lower intelligence than other children.
    b. have hearing or vision problems that should be corrected.
    c. are unmotivated and usually disruptive.
    d. have no apparent problems in hearing, vision, intelligence, or motivation.

13. The 5-to-7 shift refers to:

    a. the span of years during which there is a gradual transition from pre-operational to concrete operational thought.

    b. the sudden and seemingly overnight transition from prelogical to logical thinking.

    c. a shift from egocentric thought to perspective-taking.

    d. a shift from concrete to formal operational thought.

14. Which type of memory is as well developed in 6-year-old children as in adults?

    a. Metamemory

    b. Sensory register

    c. Long-term memory

    d. Mnemonic memory

15. The child who changes from using one form of speech to another when moving from the classroom to the playground is using:

    a. nonstandard English.

    b. the elaborated code.

    c. code-switching.

    d. colloquialism.

LESSON 12 EXERCISE: PREOPERATIONAL AND CONCRETE OPERATIONAL THOUGHT

According to Piaget, preoperational and concrete operational children think about the world in very different ways. The preoperational child (4 to 5 year-old) sees the world from his or her own perspective (**egocentrism**), and has not yet mastered the principle of **conservation**: the idea that properties such as mass, volume, and number remain the same despite changes in appearance. The concrete operational thinking of 6- to 11-year-old children is less egocentric and demonstrates mastery of logical thought, including conservation, with tangible objects.

These Piagetian concepts can be demonstrated if you know a 4- or 5-year-old and a 7- and 8-year-old—perhaps relatives, friends, or neighbors—who are willing to participate. First try several of the conservation tasks described in the audio program and in the text, p. 201. Choose from the tests for the conservation of liquid, number, matter, length, volume, and area.

Then probe your subjects' ability to take another person's point of view. Asked why the sun shines, the preoperational child might answer, "So that I can see." Try asking your subjects the following questions and any others that you can think of. Why does the sun shine? Why is there snow? Why does it rain? Also have your subjects shut their eyes, then ask if they think that you can still see them. The preoperational child is likely to say no. As Professor Kotre did in the program, inquire how many brothers and sisters each child has. Follow up by asking how many children her or his parents have. The preoperational child is likely to know the number of siblings but not the number of children his or her parents have.

After you have completed your tests, fill in the **Exercise Response Sheet** on the next page and return the sheet to your instructor.

LESSON GUIDELINES

Audio Question Guidelines

1. Preoperational children possess the notion of **phenomenalistic causality,** mistakenly believing that two things that happen together are related causally.

   Preoperational thought is also **egocentric.** Young children find it difficult to take the perspective of someone else. Egocentrism is lost gradually as the preschooler learns to keep track of two dimensions at once.

2. **Conservation** refers to the principle that an entity remains the same despite changes in its appearance. Preoperational children fail tests for conservation of volume, number, and area.

   In the test for conservation of volume, the same amount of water is poured into a short fat glass and a tall skinny glass. Failing to conserve, the preoperational child judges that the tall skinny glass contains more water.

   The failure to conserve is due to the child's failure to realize that a change in one dimension of an object brings about a change in another dimension too. Preoperational children **center** their attention on the height of the water and forget about the width of the glass.

   Children who are able to conserve take both dimensions into consideration. They have entered the stage of **concrete operations.**

3. Around the age of 6 or 7 children enter the stage of concrete operations. Now, logical operations—the workings of reason—are evident in their handling of concrete objects.

   The newly found abilities include mastery of conservation, and the emergence of **classification:** the ability to sort objects into categories and subcategories.

   The logic of concrete operational thinking means that children are also able to deal with rules. Rule-regulated thinking is evident in children's symbolic play and the school curriculum (math rules, reading rules, science rules, etc.).

   A third characteristic of concrete operational thinking is the loss of egocentrism.

   Although children are able to think logically, this is true only for tangible objects. They are not yet capable of reasoning in the abstract.

4. The preoperational child's notion of cause and effect and egocentrism have profound psychological effects. Their concepts of birth and death, for example, reflect the absence of any biological awareness of where babies come from or of the finality of death. Their sense that co-occurring events cause one another may lead them to believe that some particular act of theirs may have been the cause of their parents' divorce.

There are many social ramifications of the attainment of concrete operations, including the newly found concern with rules in symbolic play and games, and the emerging ability to take the perspective of another person.

5. Although Piaget believed that children acquire the ability to perform concrete operations around 6 or 7, contemporary researchers have found otherwise. When experiments on conservation, classification, and egocentrism are simplified, younger children are often able to respond correctly.

Other research has shown that the transition from preoperational thought to concrete operational thought is not as sudden or abrupt as Piaget believed.

Professor Elkind notes that society's concept of the child is constructed to meet the needs of adults. During the past few decades, as adults have become more "liberated," the concept of the "competent child" has emerged. As a result many parents try too hard to hurry their children's development.

Professor Elkind notes that there is no evidence that the stages of cognitive development can be accelerated, and that indeed it may be a mistake to hurry children along the path of development.

Textbook Question Guidelines

1. During the **5-to-7 shift** there is a transition from preoperational thought to concrete operational thinking.

   Until the age of 7, however, children have not quite outgrown preoperational thought nor firmly attained concrete operational abilities. A 6-year-old, for example, may correctly answer a conservation problem yet be unable to explain why the answer is correct.

   Children become increasingly able to **decenter,** are able to think more objectively, and can imagine a scene from several perspectives other than their own.

2. Many researchers believe that as a result of differences in cultural, hereditary, and educational backgrounds, cognitive development in children is more heterogeneous than Piaget's theory suggests.

   Some researchers argue that many children demonstrate signs of concrete operational thought at an earlier age than predicted by Piaget.

   Regarding school-age children, three Piagetian ideas have received consistent support: (1) compared to that of preschoolers, the thinking of school-age children is characterized by a more comprehensive logic and broader grasp of rational thought; (2) children learn best in an active mode, by questioning, exploring, and doing; and (3) *how* children think is as important as *what* they know.

3. The information-processing view of cognitive development puts more emphasis on ways that children can be taught to process the input they receive.

   School-age children become better able to receive, store, and organize information as memory improves.

   Two aspects of memory are **memory capacity** and **metamemory,** the latter involving the understanding and use of memory techniques. Memory capacity involves three levels of storage: the **sensory register,** which holds sensory images for about one second; **short-term memory,** which lasts for up to a minute; and **long-term memory,** where information can remain indefinitely.

   Adults remember better than children because they have better developed metamemory, including **selective attention, rehearsal, mnemonic devices,** and **chunking.**

   The information-processing approach emphasizes that the most effective way to teach is to adapt teaching materials and sequence of instruction to fit the needs of the individual child.

4. During the play years children's language use continues to improve in vocabulary, grammar, and pragmatics.

   School-age children begin to enjoy words as words, as demonstrated in the poems they enjoy and the jokes they tell.

   Improved understanding of logical relations helps in the understanding of comparatives (wider, deeper), of the subjunctive "if," and of metaphors.

   As egocentrism declines, communication skills and understanding of pragmatics improve.

   School-age children can easily engage in **code-switching,** from the **elaborated code** used with teachers to the **restricted code** used with friends.

5. **Specific learning disabilities** are said to occur when a child has no apparent problems in vision, hearing, intelligence, or motivation, but nonetheless has unusual difficulty learning certain skills, such as reading (**dyslexia**), writing (**dysgraphia**), or math (**dyscalcula**).

   The causes of learning difficulties are difficult to pinpoint. They may be attributable to brain damage, an inherited difficulty in brain functioning, or exposure to a particular teratogen.

   The **attention deficit disorder** manifests itself in the 5 percent of all school-age children who are overly active, impulsive, distractive, excitable. This disorder has been associated with several factors, including prenatal damage from a teratogen such as alcohol, lead poisoning, dietary deficiencies, and a restrictive upbringing.

   Although the kind of help that such children need is controversial, there are several options, including drug therapy, psychological therapy, and special classroom environments.

Answers to Testing Yourself

1. **b.** "Egocentrism" is self-defining. Thought is centered (centrism) on the self, or ego. (audio program)

2. **d.** When Piaget's original experiments are simplified, younger children are remarkably successful at them. (audio program)

3. **a.** In the conservation of liquid experiment, younger children center their attention on the height of the water and do not consider the width of the glass. (audio program)

4. **c.** Attainment of conservation represents the transition from preoperational to concrete operational thinking. (audio program)

5. **c.** Using simpler tasks for measuring conservation, classification, and egocentrism, researchers have found that younger children are able to respond correctly. (audio program)

6. **d.** Between ages 6 and 11, children can reason logically, but only about tangible things in their world. (textbook, p. 262)

7. **b.** Sorting objects by length is based on the concept of seriation, which usually does not become firmly established until age 7 or 8. (textbook, p. 264)

8. **a.** The information-processing view is so named because some aspects of human thinking are similar to the functioning of computers. (textbook, pp. 267–272)

9. **b.** Class inclusion is the concept that objects or persons may belong to more than one category or class. (textbook, p. 263)

10. **a.** The main reason why adults remember better than children is the difference in their respective metamemories. (textbook, p. 269)

11. **c.** Black English used to be considered simply poor English, until linguists realized that the so-called errors were actually consistent grammatical forms. (textbook, p. 277)

12. **d.** A child who is of average or better intelligence, yet about two years behind a specific area, may well have a specific learning disability. (textbook, pp. 280–282)

13. **a.** The 5-to-7 shift is a transitional period in which the child has not quite outgrown preoperational thought nor firmly reached concrete operational thought. (textbook p. 262)

14. **b.** Sensory register briefly stores information for less than a second and has about the same capacity in children and adults. (textbook, p. 268)

15. **c.** Children engage in many forms of code-switching, from censoring profanity when talking to parents to the bilingual child's complete switch from one language to another. (textbook, p. 276)

References

Elkind, David. (1974). *Children and adolescents: Interpretive essays on Jean Piaget.* New York: Oxford University Press.

Elkind, David. (1981). *The hurried child: Growing up too fast too soon*. Reading, MA: Addison-Wesley.

Professor Elkind presents an illuminating discussion of Piaget's theory of cognitive development and elaborates on an issue introduced in the audio program: Do adults do children a disservice by attempting to accelerate their cognitive development?

The School Years: Psychosocial Development

AUDIO PROGRAM: The First Day

ORIENTATION

This lesson on psychosocial development brings to a close the unit on the school years. Lessons 11 and 12 noted that from ages 7 to 11 children become stronger and more competent as they master the physical and cognitive skills important in their cultures. Their psychosocial development during these years is no less impressive.

As described in Chapter 13 of *The Developing Person Through the Life Span, 2/e,* the major theories of development emphasize similar characteristics in describing the school-age child. They portray an individual much more independent of the family (as stressed by psychoanalytic theory); more open to conditioning in the environment (learning theory); and better able to understand the laws and processes of the social world (cognitive theory).

Although the expanded social world of children in the school years is full of opportunities for growth, it also presents challenges and potential problems. Chapter 13 discussed the impact of low socioeconomic status, divorce, single parenthood, and maternal employment on children's psychosocial development. Most children, however, are sufficiently resilient and resourceful to cope with the stresses they may face during middle childhood. The emotional stability of parents and the amount of attention each child receives are significant factors in their healthy adjustment to environmental stress.

Audio program 13, "The First Day," focuses on the universal experience of children taking their first step onto a new stage of life: a stage where the world outside the family becomes very important. As children begin school, they confront a much more complex social world than they have previously experienced. It is a world with a new authority figure: the teacher. It is also a world with a large set of peers where there is opportunity for conversation, for play, for exploration, and the shared joy of friendship. Through the expert commentary of pychologists Steven Asher and Sheldon White, we discover both the problems and the promise of this world of school and friends and its landmark status in the life story of the developing person.

As the program opens, the host speaks with 5-year-old Hannah who, like the main character in her favorite book, is about to embark on a journey beyond the security of home and family. In Hannah's case, the journey into the world of school and friends begins with her first day of kindergarten.

LESSON GOALS

By the end of this lesson you should be prepared to:

1. Summarize three theories about the psychosocial development of school-age children.

2. Discuss the impact that peers and the social environment have on psychosocial development during middle childhood.

3. Describe the problems that may cause stress in middle childhood and factors that help to alleviate the effects of stress.

Audio Assignment

Listen to the audio tape that accompanies Lesson 13: "The First Day."
Write answers to the following questions. You may replay portions of the program if you need to refresh your memory. Answer guidelines may be found in the Lesson Guidelines section at the end of this chapter.

1. Summarize the cognitive and social skills of the 5- to 7-year-old child and describe the important transition that takes place at this age.

2. Discuss the important role that friends play in the psychosocial development of school-age children.

3. Discuss the plight of the rejected child during the school years.

Textbook Assignment

Read Chapter 13, "The School Years: Psychosocial Development," pages 291–316 in *The Developing Person Through the Life Span, 2/e.*
Write answers to the following questions. Refer back to the textbook, if necessary. Answer guidelines may be found in the Lesson Guidelines section at the end of this chapter.

1. Compare and contrast three theories about the psychosocial development of school-age children.

2. Discuss the major developments in social cognition that occur during the school years.

3. Cite examples of the "society of children" and discuss its importance.

4. Discuss the impact of their socioeconomic status on the psychosocial and academic development of school-age children.

5. Discuss the impact of family structure on psychosocial development during middle childhood.

6. Discuss the importance of competence and social support in helping children to cope with environmental stress.

Testing Yourself

After you have completed the audio and text review questions, see how well you do on the following quiz. Correct answers, with text and audio references, may be found at the end of this chapter.

1. By age 5, the preschooler's brain has attained approximately _____ percent of its eventual adult size.
   a. 65
   b. 75
   c. 85
   d. 90

2. The transition from the family environment of a young child into the broader world beyond the immediate family:
   a. occurs in children between 5 and 7 years of age throughout the world.
   b. is characteristic only of well-educated, pluralistic societies.
   c. occurs at different ages in different societies.
   d. reflects a relatively recent setting of the social clock.

3. The "vision quest" refers to:
   a. a Native American rite of passage in which young children make a journey to seek their totemic animal.
   b. a friendship game played by children the world over.
   c. the efforts of rejected children to make friends.
   d. the developmental process by which school-age children attain a sense of industry.

4. Approximately _____ percent of school children experience serious difficulty in their relationships with peers.
   a. 5
   b. 10 to 15
   c. 20 to 40
   d. 50

5. Most experts believe that the major reason school-age children are socially rejected is that they:
   a. are overly aggressive.
   b. lack positive social skills.
   c. are shy and withdrawn.
   d. are physically different.

6. According to Freud's theory, school-age children are in a period of:
   a. latency.
   b. sublimation.
   c. resistance.
   d. identity confusion.

7. According to Erikson, the crisis of middle childhood is:
   a. basic trust versus mistrust.
   b. initiative versus guilt.
   c. industry versus inferiority.
   d. identity versus role confusion.

8. Compared with preschoolers, school-age children are:
   a. more competent.
   b. less egocentric.
   c. less likely to cling to sex stereotypes.
   d. all of the above.

9. According to learning theory, reinforcement is more effective with school-age children than with preschoolers because older children:
   a. show a better understanding of cause-and-effect relationships.
   b. have a more completely developed nervous system.
   c. are more responsive to tangible rewards than younger children.
   d. show a greater desire and ability to please others.

10. When children's past failures cause them to believe that they are unable to do anything to improve their performance, they are experiencing:
    a. rejection.
    b. learned helplessness.
    c. SES.
    d. incompetence.

11. A crucial factor in attaining positive self-esteem is:
    a. having a large circle of friends.
    b. having caring parents.
    c. feeling that one is competent at varying tasks.
    d. being the recipient of prosocial behaviors.

12. Children are likely to have the lowest self-esteem at age:
    a. 5.
    b. 7.
    c. 10.
    d. 12.

13. During the school years, the influence of a child's _____ increases, while the influence of his or her _____ decreases.
    a. peers . . . parents
    b. parents . . . peers
    c. peers . . . teachers
    d. parents . . . teachers

14. During the school years, sex differences in the way children act and dress:
    a. become less distinct.
    b. become more distinct.
    c. are directly related to socioeconomic status.
    d. are primarily modeled from their parents' behavior.

15. Two factors that are especially important in helping children cope with environmental stress are:
    a. intelligence and high self-esteem.
    b. social support and a creative outlet.
    c. competence in any area and social support.
    d. high SES and competence in any area.

LESSON 13 EXERCISE: SCHOOL-AGE FRIENDSHIPS

A central theme of Lesson 13 is that middle childhood is a developmental period characterized by the child's growing inclusion in the wider social world beyond the family. This world is a complex social environment that includes a large set of peers at school and the shared joy of a smaller network of friends.

To help you apply the material from this lesson to your own life experiences, the student exercise asks you to recall the social organization of your own elementary school days and to reflect on the importance of your own friendships. You may, of course, ask the questions of someone other than yourself.

After you have completed the following questions, return your **Exercise Response Sheet** to your instructor.

LESSON GUIDELINES

Audio Question Guidelines

1. In children between 5 and 7 years of age, the biological and social clocks are in sync, preparing the child to take his or her first step into the world of school and friends.

    By age 5 the brain has reached 90 percent of its eventual adult size, children have acquired the basic vocabulary and grammar of their language, and they are on the threshold of reaching the "age of reason."

    Socially, children are able to be away from their families for relatively long periods of time, and they are able to get along with other children, forming friendships, alliances, and other social groupings.

    Among some Native American tribes, children mark this transition with the "vision quest," a journey in which they go alone into the woods to search for a totem animal that will remain associated with them for the rest of their lives. As children become connected to nature, they also become connected to the world outside their families.

2. An important task that children face during the school years is making friends.

    As children grow older, their friendships become more important, more intense, and more intimate.

    School-age children become more choosy about their friends and demand more of a smaller network of friends.

    The way that children think and talk about friendship changes as they get older. Younger children describe a friend in terms of the things they do together. Older children stress the importance of the help and emotional support their friends provide.

    Forming friendships can be very important to the child's psychological development. Children confide in their friends, and their friends help to reassure them about fears or feelings of inadequacy that they might have.

    Friends offer comfort and security to a child trying a new task or entering a new situation.

    Children also learn new skills from their friends.

3. All children experience being rejected at some time.

    An estimated 10 to 15 percent of all schoolchildren experience serious difficulty in their peer relationships.

    Such children can be grouped into several categories: aggressive children; those who are shy, withdrawn, and have feelings that are easily hurt; and those who simply lack positive social skills.

    Rejected children typically are lonely, immature in social cognition, and have low self-esteem.

    Children who have poor peer relationships in elementary school are at a greater risk of having problems coping in later life.

Teaching rejected children social skills may help them to form stable friendships.

Children should also be encouraged to develop their own interests and confidence apart from the social scene.

Research has also shown that children who improve their academic skills are likely to improve in self-esteem as well.

Textbook Question Guidelines

1. According to Freud, middle childhood is a period of **latency,** during which emotional drives are much quieter and steadier. This relative calm frees up psychic energy and allows children to put their efforts into psychosocial development.

   Erikson describes middle-childhood as a time of **industry vs. inferiority,** when children either develop competence in the skills valued by their culture or come to view themselves as inferior and inadequate.

   Learning theory notes that school-age children are particularly susceptible to the laws of conditioning.

   Operant reinforcers take on added power because children are better able to understand cause-and-effect relationships.

   Children become receptive to a wider range of reinforcements, including intrinsic rewards such as pride.

   Compared to preschoolers, school-age children model their behavior on a wider circle of people.

   Cognitive theory emphasizes the enhanced learning abilities of school-age children and their growing awareness of other points of view.

   School-age children are more logical, less egocentric, and more concrete in their thinking than younger children are.

2. **Social cognition** refers to an individual's understanding of the dynamics of human interaction.

   During the years between ages 6 and 11:

   > Children become more aware of the multiplicity of social roles an individual can have.

   > Children become aware of personality traits in themselves and others, and can anticipate how these will affect behavior.

   > Children become much better able to get along with other people.

   > Children's **self-theory** develops, enabling them to view themselves more critically.

   > Unlike younger children, school-age children tend to have lower self-esteem if they are aware of their shortcomings in any area.

   > A child's persistent failure may lead to **learned helplessness,** and a feeling of being unable to improve.

The development of positive self-esteem requires that the school-age child feel competent at varying tasks; such feelings are facilitated by a supportive microsystem of school and family.

3. The most influential system in the establishment of self-esteem is the peer group.

   Children in middle childhood have their own subculture (**society of children**), which has its own vocabulary, dress codes, and rules of behavior.

   Sex differences in clothes, behavior, and play become increasingly apparent.

   School-age children often organize clubs or gangs to distance themselves further from adults. Clubs may foster self-esteem, cooperation, and social skill development.

   In addition to developing their own social codes and society, children in middle childhood also become more aware of the customs and principles of the larger society.

4. An individual's **socioeconomic status,** measured by a combination of the education level, income, and employment status of the head of the household, is one of the most powerful influences on his or her life.

   At every stage of life, people with low socioeconomic status have a higher risk of developmental problems. These include:
   Prenatal problems, such as the mother's use of drugs.
   A greater likelihood of being neglected, abused, or accidentally hurt in childhood.
   Lower self-esteem beginning in middle childhood.
   Being more likely to drop out of school, use illicit drugs, be unemployed, and get into trouble with the law.

5. The nature of family interaction is a better predictor of problems in children than are such stereotyped characteristics of family structure as "broken home," "stepfamily," "working mother," and "father-absent" households.

   Factors that affect the child's adjustment to divorce include the amount of bitterness in the family following the divorce, the extent of changes in the child's life caused by it, the age and sex of the child, and the long-term involvement of both parents with the child after the divorce.

   The more stress parents experience, the harder it is for them to be responsive to their children. Most single parents have many sources of stress in their lives, including "role overload" and financial difficulties.

   Ecological factors, especially social support, make the difference between single-parent households that function well and those that do not.

   The effects of blended families on children vary from family to family. If the children are relatively young and living with their mother at the time she remarries, their lives typically improve, usually due to enhanced financial security for the family and higher self-esteem for the mother.
   Boys benefit particularly when their stepfathers have active and helpful relationships with them.

Maternal employment may benefit children by relieving financial pressure, and by increasing the mother's satisfaction with her life and the father's direct involvement in the household. Children with working mothers may also learn greater responsibility, which increases their competence and self-esteem.

Studies of "latchkey" children, who care for themselves after school, have shown that their self-esteem, social adjustment, and school achievement are just as good as that of children who have some form of adult care after school.

6. Variables that are associated with childhood psychiatric disorders include severe marital discord, low social status, overcrowding or large family size, paternal criminality, maternal psychiatric disorder, and admission of the child into the care of local authorities.

Children who are able to overcome several of these stressors are often referred to as "invulnerable" or "stress-resistant."

Competence in any one area and a web of social support are important in helping children to cope with stress.

Research has shown that children in school environments that expect and encourage competence are more successful in overcoming shortcomings in their community or home environment.

The role of social support is highlighted in ethnic groups in which high-risk conditions are common. Nearly half of all black American children live below the poverty level and in single-parent households. Black children benefit, however, from the fact that many black families have extensive networks of family support, including grandparents, aunts, uncles, older siblings, and neighbors.

Answers To Testing Yourself

1. **d.** By age 5, the child's brain has reached approximately 90 percent of its eventual adult size. (audio program)

2. **a.** The transition from immediate family to the larger social world is a universal phenomenon. (audio program)

3. **a.** In the vision quest, young Native American children search for their totemic animal, marking their connection with nature and the larger world outside the family. (audio program)

4. **b.** All children experience rejection at some time, but about 10 to 15 percent experience serious social difficulty. (audio program)

5. **b.** Some rejected children are aggressive or withdrawn, but all lack positive social skills. (audio program)

6. **a.** According to Freud, middle childhood is the period of latency, during which emotional drives are much quieter and steadier. (textbook, p. 291)

7. **c.** According to Erikson, as children try to master whatever skills are valued in their culture, they develop views of themselves as either industrious and productive or inferior and inadequate. (textbook p. 291)

8. **d.** During middle childhood children are more competent, less egocentric, and less likely to cling to sex stereotypes. (textbook, pp. 294–295)

9. **a.** The use of reinforcers takes on added power because school-age children are better able to understand the relationship between cause and effect. (textbook, p. 292)

10. **b.** Learned helplessness occurs when past failures in a particular area teach children to believe that they are unable to do anything to improve their performance. (textbook, p. 297)

11. **c.** Feeling that one is competent is central to the development of positive self-esteem. (textbook, p. 297)

12. **d.** Self-esteem, which is usually quite high in early childhood, decreases throughout middle childhood, reaching a low at about age 12, before it gradually rises again. (textbook, p. 296)

13. **a.** During middle childhood, children become increasingly dependent on their peers and less dependent on their parents. (textbook, p. 299)

14. **b.** Sex differences in clothes, behavior, and play patterns become increasingly salient as children move from kindergarten to the sixth grade. (textbook, p. 299)

15. **c.** Competence and social support are two important factors that help children to cope with environmental stress. (textbook, pp. 314–315)

References

Asher, Steven R., and Gottman, John M. (1981). *The development of children's friendships.* Cambridge, England: Cambridge University Press.

Various issues in the importance of friendship to psychosocial development are discussed in this collection of articles edited by Professor Asher, who is heard in the audio program.

Adolescence: Physical Development

AUDIO PROGRAM: **Changing Bodies, Changing Selves**

ORIENTATION

For most people, adolescence is an eventful season that brings dramatic changes in the ticking of each of the three developmental clocks—biological, psychological, and social. Along with physical growth that occurs at a more rapid rate than at any time since early childhood, the changes associated with sexual maturation contribute a new dimension to the ways in which adolescents think about themselves and relate to others. Lesson 14, which focuses on the nature and consequences of physical development during adolescence, begins a three-lesson unit on development between the ages of 10 and 20—the season when young people cross the boundary between childhood and adulthood.

Chapter 14 of *The Developing Person Through the Life Span, 2/e* takes a detailed look at the physical metamorphosis adolescents experience in puberty and explores such issues as adolescent dietary needs, eating disorders, and the effects of early and late maturation. It also takes a critical look at the traditional view of adolescence as a "stormy decade," noting that family, school, and cultural contexts are important factors in a young person's adjustment to puberty.

In audio program 14, "Changing Bodies, Changing Selves," the mechanisms by which the biological clock programs puberty are discussed. In this century, children are entering puberty earlier and earlier. Around the age of 9 in girls and 10 in boys, puberty begins when the pituitary gland in the brain releases hormones that affect the ovaries in girls and the testes in boys. Dr. Inese Beitins, a pediatric endocrinologist, explains the intricate sequence of bodily changes triggered by this hormonal surge. In boys and girls, the sequences are nearly opposite, with fertility arriving late in the pubertal cycle of girls and early in that of boys. The possible explanations for this sex difference are explored by anthropologists Jane Lancaster and Barry Bogin, and psychologist Laurence Steinberg. According to one view, the male-female difference was shaped by our evolutionary past, at a time when the biological and social clocks were in sync. Today, however, fertility often arrives a decade before society deems a young person to have entered adulthood, bringing with it a time when the biological clock says "ready," but the social clock says "wait."

For all of us, the changes of adolescence have a memorable and lifelong impact on our bodies and self-images. Perhaps we can recall the excitement and embarrassment of an event like the one that opens the program.

LESSON GOALS

By the end of this lesson you should be prepared to:

1. Outline the biological changes of puberty.

2. Compare the similarities and differences in male and female development during adolescence.

3. Discuss the evolutionary perspective on male-female differences in development during puberty.

4. Describe the possible problems faced by boys and girls during adolescence.

Audio Assignment

Listen to the audio tape that accompanies Lesson 14: "Changing Bodies, Changing Selves."

Write answers to the following questions. You may replay portions of the program if you need to refresh your memory. Answer guidelines may be found in the Lesson Guidelines section at the end of this chapter.

1. Outline what is known of the neural and hormonal mechanisms that govern the onset of puberty.

2. Explain why, according to some anthropologists, the biological clock programs fertility to arrive late in the pubertal sequence for girls, and early in the sequence for boys.

3. Contrast the effects of early and late maturation on boys and girls in today's society.

Textbook Assignment

Read Chapter 14: "Adolescence: Physical Development," pages 321–337 in *The Developing Person Through the Life Span, 2/e.*

Write your answers to the following questions. Refer back to the textbook, if necessary. Answer guidelines may be found in the Lesson Guidelines section at the end of this chapter.

1. Describe the sequence of physical growth during adolescence in boys and girls.

2. Discuss the nutritional needs and problems of adolescents.

3. Discuss the development of the primary and secondary sex characteristics in boys and girls during puberty.

4. Cite several factors that influence the onset of puberty.

5. Compare and contrast the traditional psychological view of adolescence with findings based on recent research.

6. Discuss research related to the long-term effects of early and late maturation, and factors related to a young person's adjustment to being physically "off-time."

Testing Yourself

After you have completed the audio and text review questions, see how well you do on the following quiz. Correct answers, with text and audio references, may be found at the end of this chapter.

1. Puberty begins at about the age of _____ in girls and _____ in boys.
   a. 10 . . . 12
   b. 9 . . . 10
   c. 11 . . . 10
   d. 13 . . . 14

2. Which of the following is true regarding the pubertal sequence of physical changes in boys and girls?
   a. For both boys and girls, fertility comes very early in the pubertal sequence.
   b. Fertility comes much earlier in the pubertal sequence for girls than for boys.
   c. Fertility comes much earlier in the pubertal sequence for boys than for girls.
   d. For both boys and girls, fertility comes very late in the pubertal sequence.

3. Compared to the turn of the century, the average age at which children enter puberty today is:
   a. younger.
   b. older.
   c. no different.
   d. less predictable and more dependent on factors such as diet and exercise.

4. One evolutionary perspective on sex differences in pubertal development holds that for our ancestors:
   a. fertile males who retained boyish features for a while were more successful reproductively than males who did not.
   b. females who did not become fertile until after they were physically developed were more successful reproductively than those who became fertile earlier.

c. the biological clock programmed fertility to occur at about the same time in males and females, with the bodily changes of puberty beginning about one year earlier in females.

d. all of the above statements are true.

5. Regarding the effects of early and late maturation on boys and girls, which of the following is *not* true?

a. Early maturation is usually easier for boys to manage than it is for girls.

b. Late maturation is usually easier for girls to manage than it is for boys.

c. Early-maturing girls may be drawn into older peer groups and may become involved in problem behaviors such as drug use and early sexual activity.

d. Late-maturing boys do not "catch up" physically, or in terms of their self-images, for many years.

6. Unlike physical development at other times during the life span, during the growth spurt of adolescence physical development is:

a. proximo-distal.

b. distal-proximal.

c. cephalo-caudal.

d. caudal-cephalic.

7. During adolescence, nutritional deficiences are most often caused by:

a. unbalanced diets.

b. caloric deficiencies.

c. anorexia nervosa.

d. insufficient protein.

8. For girls, the specific event that is taken to indicate fertility is _____; for boys it is _____

a. the appearance of breast buds . . . voice deepening.

b. dysmenorrhea . . . pubic hair.

c. menarche . . . ejaculation.

d. anovulation . . . the testosterone surge.

9. Which of the following students is likely to be the most popular in a seventh-grade class?

a. Vicki, the most sexually mature girl in the class.

b. Sandra, who is on the school debating team.

c. Brad, who is at the top of the class scholastically.

d. Dan, the tallest boy in the class.

10. Nonreproductive sexual characteristics, such as the deepening of the adolescent boy's voice and development of breasts, are called:

a. gender-typed traits.

b. primary sex characteristics.

c. secondary sex characteristics.

d. pubertal prototypes.

11. The "secular trend" refers to:
   a. the tendency over the past hundred years for puberty to come earlier and earlier.
   b. the increasing incidence of behavior problems such as drug use and early sexuality in adolescents.
   c. a decline in moral values that some developmentalists believe is occurring in our society.
   d. the effects of advertising on adolescents' body images.

12. The most significant hormonal changes of puberty include an increase of _____ in _____ and an increase of _____ in _____.
   a. estrogen . . . girls. . . . testosterone . . . boys.
   b. estrogen . . . boys . . . testosterone . . . girls.
   c. progesterone . . . girls . . . estrogen . . . boys.
   d. progesterone . . . boys . . . estrogen . . . girls.

13. The nutritional problem in which a person may rapidly lose 30 percent or more of body weight is called:
   a. bulimia nervosa.
   b. dysmenorrhea.
   c. anorexia nervosa.
   d. kwashiorkor.

14. Puberty is initiated when hormones are released from the _____, then from the _____, and then from the _____
   a. hypothalamus . . . pituitary . . . gonads.
   b. pituitary . . . gonads . . . hypothalamus.
   c. gonads . . . pituitary . . . hypothalamus.
   d. pituitary . . . hypothalamus . . . gonads.

15. Of the following, which is the most accurate general description of the period of adolescence?
   a. A period of moodiness, rebellion, and stress
   b. A period during which most individuals are well-adjusted most of the time
   c. A period when people tend to have weak or underdeveloped egos and defense mechanisms
   d. A period of turmoil caused by the hormonal changes of puberty

LESSON 14 EXERCISE: BODY IMAGES IN ADOLESCENCE

A major theme of Lesson 14 is that the physical changes of puberty have a profound impact on our self-images. Most people are able to remember at least one event, attitude, misconception, or worry they experienced in connection with the physical changes of puberty: having had big feet; having been the first or the last to experience menarche; having been concerned about having a small penis or being oversexed; having worried about acne, or voice change.

The young people who have the most difficulty are those who must adjust to these changes earlier or later than the majority of their peers. Early or late maturation may be difficult because one of the things an adolescent does not want to do is stand out from the crowd in a way that is not admirable. Early-maturing girls who are taller and more developed are often teased about their bodies by boys and accused of being "boy crazy" by girls. Late-maturing boys may compensate for being physically outdistanced by peers by becoming the class "brain," clown, or trouble-maker.

Several studies have found that early-maturing girls and late-maturing boys may experience problems that reflect their difficulty adjusting to their bodies. Early-maturing girls may be drawn to an older peer group and be more likely to engage in problem behaviors, such as drug use and early sexual activity. Follow-up studies of late-maturing boys during adulthood found that although most of them had reached average or above average height, some of the personality patterns of their adolescence persisted. Compared to early- or average-maturing boys, men who had been late-maturers tended to be less controlled, less responsible, and still have feelings of inferiority or rejection.

To help you recall your own adolescent preoccupation with appearance, the Exercise for Lesson 14 asks you to respond to questions about your body image during your own adolescence. Please answer the following questions and return your completed **Exercise Response Sheet** to your instructor. If you prefer, you may base your answers to the questions on the experiences of a friend or relative.

LESSON GUIDELINES

Audio Question Guidelines

1. The term **puberty** refers to the set of biological changes that occur at the start of adolescence and result in sexual maturity.

   Although infants have high levels of sex hormones, an unknown biological switch lowers the hormone level by the age of two and slows children's sexual development.

   At about the age of 9 in girls and 10 in boys, the **pituitary gland** in the brain releases hormones that affect the ovaries in girls and the testes in boys. In response, the ovaries and testes release sex hormones, and a year or so later the first outward signs of sexual maturation appear.

   In girls, the development of breast buds and pubic hair is followed by a growth spurt in which fat is deposited on the hips and buttocks, by the first menstrual period, and, finally, by **ovulation**—the release of the first egg.

   In boys, the normal sequence of events at puberty is nearly the reverse. First, the sex organs grow larger and pubic hair appears; then comes the first **ejaculation** of semen; then the growth spurt; and, finally, a lowering of the voice and appearance of facial hair.

2. Although girls start the sequence of puberty a year or so earlier than boys, both sexes become fertile at about the same time.

   Some anthropologists believe that puberty is coming earlier today than in previous eras because of the high fat and sugar levels in our diets.

   Other experts see the biological clock as being shaped by social needs. According to this view, early in human history, females who developed a mature appearance early in life may have had more opportunities to practice the various skills needed in the soon-to-be-attained social roles as adults. Females who followed this growth pattern may have been more successful in having babies and rearing more of those babies to adulthood. In this way, the female pubertal sequence may have become coded into the timing of the biological clock.

   In boys, according to this viewpoint, early development of the physical characteristics of stature, muscles, facial hair, and a deep voice might have been dangerous. If boys looked like men before they were fertile, they might have been perceived as competitors by older men. Such individuals would probably have been less successful reproductively, leading to the coding of mature appearance late in the pubertal sequence of boys' biological clocks.

3. The adolescents who have the most difficult time with puberty are early-maturing girls and late-maturing boys.

   Early-maturing girls are often teased and may suffer a loss of self-esteem.

   Early-maturing girls may be drawn into an older peer group and be more likely to engage in problem behavior, such as drinking, drug use, and early sexual activity.

Late-maturing boys are at a disadvantage because, being smaller and less muscular, they tend to be less competitive at sports—the ticket to popularity for boys in most junior and senior high schools. This may have a continuing negative impact on the late-maturing boy's self-image.

Textbook Question Guidelines

1. **Puberty** begins with increased levels of hormones in the bloodstream—**estrogen** especially in girls, and **testosterone** more markedly in boys.

   The first sign of the onset of puberty—the **growth spurt**—occurs about a year after this increased hormone release. Between ages 10 and 12, boys and girls become noticeably heavier, through the accumulation of fat on thighs, arms, buttocks, and abdomen. Between ages 10 and 14, a typical girl gains about 38 pounds and 9⅝ inches, while a typical boy gains the same in height and about 42 pounds.

   The sequence of growth in puberty is **distal-proximal** (from far to near), making many adolescents temporarily big-footed, long-legged, and short-waisted.

   Internal organs, including lungs and heart, also grow, providing the adolescent with increased physical endurance.

   The lymphoid system decreases in size during adolescence, making teenagers less susceptible than children to respiratory ailments.

2. Nutritional requirements during adolescence include a need for additional calories, protein, calcium, iron, zinc, and vitamin D. **Iron deficiency anemia,** especially in girls, is common during adolescence.

   The most serious nutritional problem of adolescence is **anorexia nervosa.**

   Anorexia is more common in girls than boys, is often linked to depression, and is often related to a disturbed body image.

   Without outside help, about ⅓ of anorexics get better and about ⅕ die.

   Psychoanalytic therapists contend that the anorexic girl is afraid of the responsibilities of becoming a woman and maintains a childlike form by extreme dieting.

   Learning theorists view eating disorders as maladaptive behaviors intended to gain social reinforcement from parents and peers.

3. Growth and maturation of the sex organs (**primary sex characteristics**) result in the development of reproductive potential, signaled by **menarche** (the first menstrual cycle) in girls and the first **ejaculation** of semen in boys.

   Full fertility is not achieved until several years later, however. The first menstrual cycles are usually anovulatory, and in boys the concentration of sperm is relatively low.

   Some adolescent girls are occasionally unable to carry on normal activities because of menstrual pain (**dysmenorrhea**).

   Development of **secondary sex characteristics**—breasts and pubic hair in girls; pubic and facial hair, and lowering of the voice in boys—contributes to the adolescent's new **body image.** In developing this image, most adolescents refer to the cultural body ideal: the slim, shapely woman and the tall, muscular man promoted by advertising. As a result of such comparisons, most adolescents are dissatisfied with their own appearance.

4. Puberty begins when hormones from the **hypothalamus** trigger the secretion of hormones in the **pituitary gland,** which increases the production of hormones—**estrogen** and **progesterone** in girls, and **testosterone** in boys—by the **gonads** (ovaries and testes).

   The onset of puberty is influenced by many factors, including the child's sex, genes, body type, nutrition, and health.

   As a result of better nutrition and health care, each new generation over the past 100 years has experienced puberty earlier than the previous generation —a tendency known as the **secular trend.**

5. The view that adolescence is inevitably a time for storm and stress was first expressed by G. Stanley Hall. Hall argued that physical maturation changes not only adolescents' size and physiology, but also their emotional and moral development.

   Psychoanalytic theorists believe that the stresses of adolescence are inescapable, since the adolescent's rapid sexual maturation and sexual drives inevitably conflict with the culture's prohibitions against sexual expression at this age.

   Contemporary studies have found, however, that most adolescents are calm and predictable most of the time.

6. Early-maturing girls and late-maturing boys tend to have the greatest difficulty adjusting to puberty.

   Early-maturing girls may date early, and suffer a loss of self-esteem, feeling constantly scrutinized by parents and friends and pressured by dates into premature sexual activity. They are also more likely to have sexual intercourse before finishing high school, less likely to use contraception, and more likely to become pregnant.

   Most early-maturing girls, however, avoid premature sexual experiences and, by seventh or eighth grade, experience increased status, respect, and popularity among their peers.

   The Berkeley Growth Study found that late-maturing boys were less poised, less relaxed, more restless, and more talkative than early-maturing boys. They also possessed characteristics not usually admired by their peers, such as being more playful, creative, and flexible.

   As adults, those who had matured later tended to be less controlled, less responsible, less dominant, and were less likely to hold positions of leadership. Some still had feelings of inferiority and rejection.

   The effects of early or late maturation are more apparent among lower-class adolescents, because physical prowess tends to be more highly valued than among middle- or upper-class teenagers.

When the family already has a pattern of dysfunction, or when stressful family events coincide with an adolescent entering puberty, he or she is more likely to have difficulty adjusting.

Answers to Testing Yourself

1. **b.** (audio program)

2. **c.** Although pubertal changes begin a year earlier for girls than for boys, fertility occurs near the end of the sequence in girls and near the beginning of the sequence in boys. (audio program)

3. **a.** For the past 100 years or so, each generation has experienced puberty at an earlier age than the preceding generation. (audio program)

4. **d.** All of these are true according to the evolutionary perspective discussed in the program. (audio program)

5. **d.** Early-maturing boys generally do catch up physically within a relatively short period of time. (audio program)

6. **b.** Unlike that in other seasons of life, growth during puberty is from far to near. (textbook, p. 323)

7. **a.** Teenagers are particularly susceptible to the latest crash diets and food fads. (textbook, p. 325)

8. **c.** Although menarche is considered a landmark, normally another year or so must pass before a girl becomes fertile. (textbook, p. 326)

9. **d.** Early-maturing boys tend to be more popular than late-maturing boys and early-maturing girls. (textbook, pp. 333–336)

10. **c.** The development of secondary sex characteristics also indicates that sexual maturation is occurring. (textbook, p. 327)

11. **a.** This trend is the consequence of good nutrition and better health generally. (textbook, p. 330)

12. **a.** Although levels of estrogen and testosterone increase somewhat in both boys and girls, the increases are most marked in girls and boys, respectively. (textbook, pp. 321–322)

13. **c.** The most common ages of onset of anorexia nervosa are about 11 and 18—the beginning and end of adolescence. (textbook, p. 325)

14. **a.** Puberty begins when hormones from the hypothalamus trigger hormone production in the pituitary, which in turn triggers increased hormone production by the gonads. (textbook, p. 328)

15. **b.** Most developmental psychologists today do not believe that adolescence necessarily conforms to the popular image of a period of storm and stress. (textbook, p. 331)

References

Katchadourian, Herant A. (1977). *The biology of adolescence*. San Francisco: WH Freeman.

A very readable and comprehensive description of the physical changes of puberty and the biological mechanisms by which they are programmed.

Adolescence: Cognitive Development

ORIENTATION

Lesson 14 explored the biological changes that transform children into adolescents by giving them adult size, shape, and reproductive capacity. But equally important psychological and social changes occur during this sometimes tumultuous season. Lesson 15, which focuses on cognitive development, is the second of a three-lesson unit on adolescent development.

During adolescence, young people become increasingly able to speculate, hypothesize, and use logic. Unlike younger children, whose thinking is tied to **concrete operations,** adolescents with the ability to think in terms of **formal operations** are able to consider possibilities as well as reality. Logical thought processes also foster **moral development,** as adolescents become better able to grasp moral laws and ethical principles. Chapter 15 of *The Developing Person Through the Life Span, 2/e,* discusses adolescent cognitive and moral development through the theories of Jean Piaget and Lawrence Kohlberg, respectively.

Adolescent thinking has its limitations, however. As psychologist David Elkind notes in audio program 15, "All Things Possible," adolescents often create for themselves an **imaginary audience** and a **personal fable,** as they fantasize about how they appear to others, imagine their own lives as heroic, and feel that they are immune to danger.

The audio program also explores a new theory of intelligence proposed by psychologist Robert Sternberg. Professor Sternberg believes that **practical intelligence** reflects an ability to function effectively in everyday life. The emergence of practical intelligence may help adolescents to develop a more realistic picture of who they are.

Perhaps most significantly, the logical, idealistic, and egocentric thinking of adolescence represents a significant step in the process of creating a life story. Such is the view of psychologist Dan McAdams, who sees this season as a time when individuals begin to define that which makes them unique. For Erik Erikson, adolescence is a time when young people struggle to form an **identity.** As they do, they create a background of **ideology** and write the first draft of their own life stories.

LESSON GOALS

By the end of this lesson you should be prepared to:

1. Describe the cognitive characteristics of the typical adolescent.

2. Discuss the cognitive component of adolescent problems with sexual behavior and drug use.

3. Identify ways in which adults can foster mature decision-making by adolescents.

4. Discuss the significance of adolescent thinking in the formation of identity and the process of creating a life story.

Audio Assignment

Listen to the audio tape that accompanies Lesson 15: "All Things Possible."
Write answers to the following questions. You may replay portions of the program if you need to refresh your memory. Answer guidelines may be found in the Lesson Guidelines section at the end of this chapter.

1. Explain how the development of formal operational thinking influences adolescents' attitudes toward family and society.

2. Compare and contrast formal operational thought with practical intelligence.

3. Discuss the significance of the imaginary audience and personal fable in the adolescent's formation of identity and creation of a life story.

Textbook Assignment

Read Chapter 15: "Adolescence: Cognitive Development," pages 339–361 in
The Developing Person Through the Life Span, 2/e.

Write answers to the following questions. Refer back to the textbook, if
necessary. Answer guidelines may be found in the Lesson Guidelines section at
the end of this chapter.

1. Summarize Piaget's description of cognitive development during adolescence.

2. Discuss adolescent egocentrism and its corresponding fantasies and fables.

3. Outline Kohlberg's stages of moral reasoning and explain how these stages
are related to cognitive development.

4. Cite several criticisms of Kohlberg's theory of moral development.

5. Identify the main issues related to adolescent sexual behavior and discuss some of the possible explanations for the high incidence of sexually transmitted disease and pregnancy among teenagers.

6. Discuss the use of drugs by adolescents and identify several reasons why adolescents may have difficulty making mature decisions regarding drug use.

Testing Yourself

After you have completed the audio and text review questions, see how well you do on the following quiz. Correct answers, with text and audio references, may be found at the end of this chapter.

1. One distinguishing feature of adolescent thinking is the ability to think in terms of:
   a. concrete operations.
   b. sensory–motor schemas.
   c. "us" versus "them."
   d. possibility as well as reality.

2. The imaginary audience refers to:
   a. the concrete operational child's preoccupation with performing for everyone, including toys and imaginary people.
   b. the adolescent's fear of being spied on.
   c. the ability of adolescents to empathize with others by putting themselves in their shoes.
   d. the egocentric fantasy of adolescents that others are constantly attending to their behavior and appearance.

3. According to Piaget, the final stage of cognitive development is the stage of:
   a. concrete operations.
   b. practical intelligence.
   c. formal operations.
   d. post-formal operations.

4. A chief executive officer of a major company who does poorly on standardized intelligence tests, yet is very successful professionally, would probably be described by Robert Sternberg as possessing considerable:
   a. practical intelligence.
   b. fluid intelligence.
   c. crystallized intelligence.
   d. egocentrism.

5. The personal fable refers to adolescents':
   a. preoccupation with what they perceive as faults in their appearance and personality.
   b. fantasy of being destined to live heroic lives of fame and fortune.
   c. belief that they are the center of attention everywhere they go.
   d. earliest memory and myths of their origin.

6. According to Kohlberg, the three levels of moral reasoning are:
   a. amoral, immoral, moral.
   b. egocentrism, ethnocentrism, universality.
   c. preconventional, conventional, postconventional.
   d. selfishness, law-and-order, altruistic.

7. Which of the following has *not* been raised as a criticism of Kohlberg's theory of moral development?
   a. The stages reflect liberal Western values.
   b. The theory is biased against females.
   c. The theory overemphasizes rational thought processes.
   d. The theory overestimates the importance of religious faith in moral reasoning.

8. According to the textbook, the sexual attitudes of adolescents today are:
   a. no different than those of adolescents 50 years ago.
   b. more dangerous than those of previous generations.
   c. healthier than those of previous generations.
   d. more confused than those of previous generations.

9. Some adolescents falsely believe that dangerous drugs will not hurt them. This belief is an expression of the:
   a. imaginary audience.
   b. game of thinking.
   c. illusion of immunity.
   d. invincibility fable.

10. Illusions of invulnerability, personal fables, and imaginary audiences are manifestations of adolescent:
    a. games of thinking.
    b. egocentrism.
    c. practical intelligence.
    d. formal operations.

11. An experimenter hides a ball in her hand and says, "Either the ball in my hand is red or it is not red." Most preadolescent children say:
   a. the statement is true.
   b. the statement is false.
   c. they cannot tell if the statement is true or false.
   d. they do not understand what the experimenter means.

12. One difficulty with Piaget's theory is that:
   a. many adults have difficulty with formal operational thinking.
   b. formal operational thinking appears only in industrialized countries, such as the United States.
   c. formal operational thinking typically emerges at a much earlier age than Piaget proposed.
   d. formal operational logic often emerges before the principle of conservation is mastered.

13. According to the textbook, the best advice for parents in dealing with adolescents' problems with drugs and sex, is to:
   a. lay down strict rules.
   b. listen, discuss, and debate.
   c. be a "best friend."
   d. do all of the above.

14. The drug most commonly used by adolescents today is:
   a. tobacco.
   b. alcohol.
   c. marijuana.
   d. cocaine.

15. Moral development is most closely related to:
   a. social development.
   b. physical development.
   c. cognitive development.
   d. academic intelligence.

LESSON 15 EXERCISE: LOGICAL VS. PRACTICAL INTELLIGENCE

During adolescence, formal operational thought—including scientific reasoning, logical construction of arguments, and critical thought—becomes possible. Piaget's finding that adolescents are more logical and systematic than preadolescents has been replicated in experiments many times. But not all problems encountered by adolescents and adults require thinking at this level. Also, while many adolescents and adults are capable of thinking logically, they do not always do so. Indeed, older adults often find that a formal approach to solving problems is unsatisfactory and oversimplified.

One theme of this lesson is that a new kind of **practical intelligence** begins to emerge during late adolescence and early adulthood. This type of thinking is more applicable to everyday situations than formal thought. It recognizes that for many problems there may be no single correct answer, and that "logical" answers are often impractical. Some researchers believe that this new way of thinking reflects the greater cognitive maturity of adults in reconciling formal thought with the reality of their lives. If this is true, measuring adult intelligence by the same standards used to assess the "pure" logic of the adolescent is clearly inappropriate.

To stimulate your own thinking about the difference between formal thought and practical intelligence, consider the following problems. After you have answered the questions, return your completed **Exercise Response Sheet** to your instructor.

LESSON GUIDELINES

Audio Questions

1. The development of formal operational thought enables adolescents to move beyond the concrete operational world of "here-and-now" to a world of possibility, where abstract, logical, and scientific thinking becomes typical.

   Because they are able to speculate about possibilities, adolescents often think about ideal parents and an ideal society. This idealism may cause some teenagers to become dissatisfied with their present situation and want to change it. Adolescents are also able to imagine ideal mates, and so they are susceptible to "crushes."

2. Formal operational thought is hypothetical, logical, and abstract. This kind of thinking is emphasized by schools and teachers.

   According to Robert Sternberg, most of the everyday problems people face do not have clear correct and incorrect solutions that can be arrived at by using formal operational thinking.

   In addition to formal operational, or academic intelligence, a valuable kind of thinking that educators generally overlook is **practical intelligence.** This is the ability to adapt, shape, and select real world environments. It represents an ability to function effectively in the everyday world.

   Many people who have relatively low scores on standardized tests of academic intelligence may nevertheless be very successful in careers in which practical intelligence is especially useful.

3. According to David Elkind, adolescent egocentrism, which is different from childhood egocentrism, leads young people to see themselves as being much more central to their social world than they really are. Adolescents often create an imaginary audience for themselves and fantasize about how others will react to their appearance and behavior.

   Egocentrism may also lead adolescents to develop their own **personal fable,** in which they imagine their lives as heroic or destined for fame and fortune, and an **invincibility fable,** in which they see themselves as immune to dangers.

   These cognitive tendencies prepare the way for the process of forming a life story. Through them the teenager begins to define how he or she is different from others, which leads to the formation of an identity. Over time, the early fantasies and fables become more realistic. The older adolescent develops a belief and value system that serves as an **ideological** background for his or her life story.

Textbook Questions

1. Piaget was the first to recognize the capacity of adolescents to think in terms of possibility rather than merely concrete reality. **Formal operational thought** permits adolescents to speculate, form hypotheses, and use **scientific reasoning.**

Another characteristic of formal operational thought is the ability to think creatively about hypothetical possibilities, using what Flavell calls the game of thinking.

Formal operational reasoning is apparent in other domains, including the development of historical reasoning, ecological judgment, and sociopolitical reasoning.

One limitation of Piaget's theory is that not all adults are able to pass tests of formal operational thought. Piaget argues that the maturation of brain and body make formal operational thought *possible,* but not inevitable. The influences of society and education are important factors in determining whether a person attains formal operational reasoning.

2. Young people often consider themselves unique, and they experience a different kind of egocentrism from that of children. **Adolescent egocentrism** permits them to create an **imaginary audience** that watches their every move, and to believe a **personal fable** that their lives are heroic and immune to the laws of mortality **(invincibility fable).** This type of thinking indicates that the adolescent is not at ease with his or her larger social world.

3. Cognitive development allows adolescents to think more abstractly, to question the moral beliefs of society, and to progress in moral reasoning.

Kohlberg identified three levels of moral reasoning, with two stages at each level.

Preconventional: Emphasis on avoiding punishment and obtaining rewards. Stage 1: punishment and obedience orientation ("Might makes right"). Stage 2: instrumental and relativist orientation ("Look out for number one").

Conventional: Emphasis on social rules. Stage 3: "good girl" and "nice boy." Stage 4: "law and order."

Postconventional: Emphasis on moral principles. Stage 5: social contract. Stage 6: universal ethical principles.

4. Kohlberg's stages may reflect liberal Western values rather than universal ethical principles.

The theory may be biased against females, who tend to see moral dilemmas differently from males. According to Carol Gilligan, males take a "do not interfere with the rights of others" approach, while females tend to be more concerned with the context in which moral choices must be made.

Kohlberg's stages may overemphasize rational thought and underestimate the importance of religious faith in moral reasoning.

A verbally fluent person may score quite well on Kohlberg's measures without really understanding moral reasoning.

Kohlberg assumed that moral development is sequential when, in fact, many people progress and regress in their level of reasoning.

5. Although adolescent attitudes toward sexuality are probably healthier today than the taboos and double standards earlier generations were subject to, the rising incidence of **sexually transmitted disease (STD)** and adolescent

pregnancy make it clear that cognitive immaturity coupled with sexual freedom can be destructive.

STD is a leading cause of infertility and, in the case of acquired immune deficiency syndrome (AIDS), illness and death.

Almost half of all teenage girls in the United States become pregnant at least once before they are 20.

Adolescents are less likely to use contraception than adults are. Boys typically regard pregnancy as primarily the girl's problem.

Adolescents often are illogical in their thinking about sexuality. Adolescent girls who have reason to believe they may be pregnant often postpone seeing a doctor. Because they do not want to be pregnant, they succumb to the invincibility fable, "It can't happen to me."

6. By their senior year of high school, almost all adolescents have used alcohol and a majority have tried tobacco and marijuana. Cocaine and alcohol use are increasing and adolescents in the 1980s are generally reporting first use of various drugs at younger ages than did adolescents in the 1970s.

Social pressures lead adolescents to try the most common drugs. The adolescent inclination to feel invulnerable may lead to the belief that they can handle drugs easily.

Adolescent drug use, as well as sexual behavior, is often influenced by the examples parents set.

Even more important than providing adolescents with the facts concerning drug use is fostering development of reasoning that allows them to apply facts to their own experience. Parents' and teachers' best strategy is to listen and debate, which encourages adolescent thought about social and moral issues.

Answers to Testing Yourself

1. **d.** Before she or he attains formal operational thought, the child is limited to thinking about the concrete world of here and now. The adolescent is capable of dealing with possibilities and with the future. (audio program)

2. **d.** The imaginary audience is an especially vivid expression of adolescent egocentrism. (audio program; textbook, p. 344)

3. **c.** (audio program; textbook, p. 339)

4. **a.** Practical intelligence is the ability to function effectively in the everyday world. (audio program)

5. **b.** The personal fable is another manifestation of adolescent egocentrism. (audio program; textbook, p. 346)

6. **c.** Kohlberg's stages of moral development are closely related to cognitive development. (textbook, pp. 347–348)

7. **d.** In fact, Kohlberg's theory has been criticized for underestimating religious faith in moral reasoning. (textbook, p. 349)

8. **c.** The decline of the double standard and the contemporary increase in sexual understanding mean that fewer of today's young people face the negative feelings about sexuality that were pervasive in recent generations. (textbook, p. 353)

9. **d.** Many young people feel that they are immune to the laws of mortality and probability; consequently, they take many risks. (audio program; textbook, p. 346)

10. **b.** Adolescents tend to see themselves as much more significant in the social environment than they actually are. (textbook, pp. 344–346)

11. **c.** Although this statement is logically verifiable, preadolescents who lack formal operational thought cannot prove or disprove it. (audio program; textbook, p. 341)

12. **a.** Many older adolescents, including college students, do poorly on standard tests of formal operational thought. (textbook, p. 342)

13. **b.** Because of adolescents' egocentrism, the best long-range strategy for adults is to listen and debate, rather than to lay down the law. (textbook, p. 360)

14. **b.** Drug use among adolescents varies from year to year and from cohort to cohort. (textbook, p. 356)

15. **c.** Once people can imagine alternative solutions to various problems in science or logic, they can begin to apply these same mental processes to thinking about right and wrong. (textbook, p. 347)

References

Sternberg, Robert J. (1986). *Intelligence applied: Understanding and increasing your intellectual skills.* New York: Harcourt Brace Jovanovich.

In this highly readable book, Professor Sternberg discusses several theories of intelligence (including those of Jean Piaget), develops his concept of practical intelligence, and explains how people can improve their intellectual skills.

Adolescence: Psychosocial Development

AUDIO PROGRAM: Second Chances

ORIENTATION

As discussed in Lessons 14 and 15, the physical changes of adolescence transform the child's body into that of an adult, while the cognitive changes enable the young person to think logically and more practically. These changes set the stage for psychosocial development, which is the subject of Lesson 16.

According to Erik Erikson, adolescence brings the dawning of commitment to a personal identity and future, to other people, and to ideologies. Friends, family, community, and culture are powerful social forces that act to help or hinder the adolescent's transition from childhood to adulthood.

Chapter 16 of *The Developing Person Through the Life Span, 2/e,* focuses on the adolescent's efforts toward **identity achievement.** The influences parents and peers have on psychosocial development are examined in detail, as are the normal difficulties and special problems of adolescence. These problems include **delinquency, sexual abuse,** and **suicide.**

Audio program 16, "Second Chances," examines the psychosocial challenges of adolescence, emphasizing the particular vulnerability of today's teenagers. This vulnerability is dramatically illustrated by the stories of two teenage "casualties." Although the stories of Valerie and Tony have troubled beginnings, both teenagers were afforded "second chances" and seem to be taking advantage of them.

Although many of today's teenagers are visibly damaged by the problems of this tumultuous season, the image of the troubled adolescent as irretrievable is incorrect. Through the expert commentary of psychologists Ruby Takanishi and Richard Jessor, we discover that although peer group pressure is often the initial trigger for problem behaviors in adolescence, the socializing role of peers has many potentially positive effects. These include facilitating identity formation, independence, and the development of social skills that help the young person eventually attain adult status and maturity.

As adolescents forge an identity and attempt to make wise choices about their futures, their social context encourages some paths to identity and forecloses others. The result of this interaction will be, in the ideal case, young people who are sure of themselves and are able to pass through these vulnerable years of adolescence successfully.

LESSON GOALS

By the end of this lesson you should be prepared to:

1. Describe the four identity statuses of adolescence.

2. Discuss the role of friends and peer groups in identity formation.

3. Describe the possible effects of various parenting styles on identity formation in adolescence.

4. Explain why early adolescence—the years from 10–14—is an especially critical time for young people today.

Audio Assignment

Listen to the audio tape that accompanies Lesson 16: "Second Chances."

Write answers to the following questions. You may replay portions of the program if you need to refresh your memory. Answer guidelines can be found in the Lesson Guidelines section at the end of this chapter.

1. Explain why early adolescence is considered a period of particular vulnerability for young people today.

2. Discuss the issue of retrievability in adolescence. Who needs a "second chance," and who is most likely to receive one?

3. Identify and explain the kinds of interventions that have helped adolescents to take advantage of such second chances.

Textbook Assignment

Read Chapter 16: "Adolescence: Psychosocial Development," pages 363–386 in *The Developing Person Through the Life Span, 2/e.*

Write answers to the following questions. Refer back to the textbook, if necessary. Answer guidelines can be found in the Lesson Guidelines section at the end of this chapter.

1. Describe Erikson's view of the development of identity in adolescence.

2. Describe four major identity statuses that are typical of adolescence.

3. Define "rite of passage" and give examples of this process for both Western and non-Western cultures.

4. Compare and contrast the influence of parents and the peer group on adolescent psychosocial development.

5. Identify seven styles of parenting and discuss their possible effects on adolescent development.

6. Identify three significant problems experienced by today's adolescents, indicating their causes and how they might be alleviated.

Testing Yourself

After you have completed the audio and text review questions, see how well you do on the following quiz. Correct answers, with text and audio references, may be found at the end of this chapter.

1. Which of the following is the most common problem behavior among adolescents?
   a. Pregnancy
   b. Daily use of addictive drugs
   c. Minor law-breaking
   d. Suicide

2. Crime statistics show that during adolescence:
   a. males and females are equally likely to be arrested.
   b. males are more likely to be arrested than females.
   c. females are more likely to be arrested than males.
   d. males commit more crimes than females, but are less likely to be arrested.

3. Regarding the influence of peer group pressure in producing unwanted adolescent behavior, which of the following is true?
   a. Pressure to conform rises dramatically in early adolescence and then declines.
   b. Pressure to conform increases throughout this season, peaking in late adolescence.
   c. Pressure to conform peaks in late childhood, and decreases throughout adolescence.
   d. The developmental trajectory of peer group influence is unpredictable.

4. As discussed in the audio program, "retrievability" refers specifically to the:
   a. cognitive capacity of adolescents to remember their own internalized ideologies.
   b. ability of troubled adolescents to bounce back from problem behaviors and benefit from "second chances."
   c. influence of autocratic parenting on adolescent identity formation.
   d. permanent mark left on identity by problem behaviors during adolescence.

5. Peer group pressure is likely to be the strongest at age:
   a. 8.
   b. 11.
   c. 14.
   d. 17.

6. According to Erik Erikson, the primary task of adolescence is:
   a. developing an intimate relationship.
   b. attaining a sense of integrity.
   c. forming an identity.
   d. achieving a sense of industry.

7. The identity status of foreclosure is one in which the individual:
   a. is very apathetic about trying to find an identity.
   b. takes a time-out in order to experiment with alternative identities.
   c. adopts an identity that is the opposite of the one he or she was expected to adopt.
   d. prematurely adopts an identity without exploring alternatives.

8. The adolescent experiencing identity diffusion is typically:
   a. very apathetic.
   b. a risk-taker anxious to experiment with alternative identities.
   c. willing to accept parental values wholesale, without exploring alternatives.
   d. one who rebels against all forms of authority.

9. Of the following, which is the best example of a rite of passage?
   a. Obtaining a driver's license
   b. Passing a difficult college course
   c. Choosing a wardrobe
   d. Making the school basketball team

10. Which of the following most accurately describes the friendship circle of a typical adolescent?
    a. A small and stable group of friends
    b. A small and even-changing group of friends
    c. A large and stable group of friends
    d. A large and ever-changing group of friends

11. The most crucial predictor of an adolescent's future achievement and mental health is his or her:
    a. parents' values.
    b. intelligence.
    c. socioeconomic status.
    d. ability to get along with peers.

12. Which styles of parenting do most psychologists recommend?
    a. Laissez-faire and permissive
    b. Autocratic and authoritarian
    c. Democratic and equalitarian
    d. Reciprocal and genuine

13. Autocratic and authoritarian patterns of parenting are common among:
    a. large families.
    b. low-income families.
    c. families with younger adolescents.
    d. all of the above.

14. Research has found that one of the best predictors of later delinquent behavior is when an adolescent:
    a. quarrels frequently with brothers and sisters.
    b. often questions parental values.
    c. experiences difficulty in school.
    d. is considered a leader in his or her peer group.

15. Which of the following was not identified as a warning sign of suicide in an adolescent?

   a. A sudden decline in school achievement
   b. An attempted suicide
   c. A break in a love relationship
   d. A sudden interest in friends and family

LESSON 16 EXERCISE: IDENTITY THROUGH THE SEASONS

A central theme of the three-lesson unit on adolescence is that identity formation is a primary task of this season. Ideally, adolescents develop a clear picture of their own uniqueness in the larger social world of which they are a part.

But the development of identity is not confined to one season of life. Like the life story itself, it continues to evolve over the life span. To help you apply this truth to your own life story, the exercise asks you to respond to the deceptively simple question, "Who am I?" You may respond in terms of your social roles, responsibilities, or commitments; the groups to which you belong; your beliefs and values; personality traits and abilities; and your needs, feelings, and behavior patterns. List only things that are really important to you—things that, if lost, would make a real difference in your sense of who you are. Limit your answer to 10 definitions of your identity.

After you have completed your list, you are asked to consider each item separately and assign a number from 1 (most important) to 10 (least important) to each item, indicating its importance to your identity today. Then rank the items again, based on their importance to you 10 years ago. What differences do you see?

As always, you may ask someone else to answer the questions on this exercise. After you have completed the assignment, return your **Exercise Response Sheet** to your instructor.

LESSON GUIDELINES

Audio Question Guidelines

1. The biological changes associated with puberty are occurring among young people today earlier than in the past. Society, however, has not adapted to this change. The result is a relatively long period of time during which young people are physically mature, yet socially denied the privileges and responsibilities of adults.

   There is great diversity in physical development during early adolescence; some individuals continue to look like children while others already resemble adults. This diversity is itself challenging for adolescents.

   Today there may be greater opportunity in the adolescent environment for experimentation and risk taking. This includes pressure from peers to experiment with risky behaviors such as delinquent acts, early sexual activity, and use of alcohol and other drugs.

   Compared to the experience of previous generations, the substances to which adolescents are exposed today are much more lethal. This may account for the fact that the leading cause of death among young adolescents is accidents, often related to the diminished judgment that accompanies the use of drugs.

   Poverty provides an additional burden for many young people.

2. Approximately 75 percent of adolescents go through this season relatively easily; about 25 percent are at risk and do not make the transition very well.

   For those in early or middle adolescence who have been disadvantaged and have already begun to experiment with risky behaviors, opportunities for "second-chance" interventions are especially important.

   Individuals who have been at risk need to perceive that they have a promising future economically and psychologically.

   Many young people have reserves of resiliency that make them eminently "retrievable" from their disadvantaged beginnings. For young people who do not have this inner tenacity, society's early provision of programs may make a major difference.

3. Peer counseling may help adolescents to learn about the dangers of high-risk behaviors, develop independence and social skills, forge a clearer sense of identity, and promote self-esteem. It can be helpful both to those who give it and those who receive it.

   Programs that link young people to consistent, caring adults have proven to be beneficial. So has the continuing involvement of parents.

Textbook Question Guidelines

1. According to Erikson, the primary task of adolescence is the search for **identity**, both as an individual and as part of the larger community.

   The ultimate goal is **identity achievement,** which occurs when adolescents develop their own ideology and vocational goals.

2. Some adolescents achieve identity prematurely, in a process called **foreclosure;** others experience aimless **identity diffusion** with few commitments to goals, values, or society.

Some simply rebel and become the opposite of what their parents want, adopting a **negative identity.**

Some adolescents declare a **moratorium** on identity formation, often by choosing an institutionalized time-out such as college or military service as a way of postponing final decisions about career or marriage.

Research studies have shown that each of the four major identity statuses relates to a unique set of characteristics regarding attitudes toward parents, self-esteem, ethnic identity, prejudice, moral development, and cognitive style.

3. In adolescence, society provides an avenue for identity formation in two ways: by providing time-tested values, and by providing social customs that ease the transition from childhood to adulthood.

    A **rite of passage** is a ceremony or event that provides a transition from one social status or life stage to another.

    In many non-Western cultures, initiation ceremonies for males were dramatic and painful tests of strength and bravery, often involving circumcision or facial scarring. Female rites were usually intended to prepare the young woman for courtship, marriage, and her role in running a household.

    Examples of rites of passage in our own society include **religious rites** (Catholic confirmation; Jewish Bar Mitzvah and Bat Mitzvah), **social rites** (debutante or sweet sixteen parties), and **legal proceedings** (voter and driver registration).

4. The peer group eases the transition from childhood to adulthood by functioning as a support group for adolescents undergoing new experiences and challenging parental values, and as a forum in which young people can discover which personality characteristics and behaviors are accepted.

    While school-age children are typically closer to their parents than to friends, as young people enter adolescence, they are more likely to share personal information with peers of both sexes than with their parents.

    The peer group also facilitates the adolescent's switch from same-sex groups to intimate relationships with the opposite sex.

    Generally speaking, peers supplement rather than undermine parents' influence. Despite the minor conflicts that many adolescents have with their parents, young people follow their parents' lead on most issues.

    The **generational stake** refers to the fact that each generation in the parent–adolescent relationship has its own particular needs and concerns, as well as a natural tendency to see the family in a certain way.

5. **Autocratic.** Young people make no decisions about their own lives.

    Authoritarian. Although the young person is permitted to contribute opinions, parents always have the final say.

    Democratic. Adolescents contribute to the discussion of relevant issues and make some of their own decisions. Final decisions are subject to parental approval.

Equalitarian. Parents and adolescents participate equally in decision-making.

Permissive. The adolescent plays the more influential role in making decisions and does not always abide by parental opinions.

Laissez-faire. Parents leave it to the young person to decide to consider or ignore parental opinions.

Ignoring. Parents take no role in directing the adolescent's behavior.

The particular style of parenting influences the development of the young person. The extremes of autocratic and authoritarian parenting on the one side, and permissive, laissez-faire, or ignoring parenting on the other, do not foster development as well as do democratic or equalitarian family interaction.

Autocratic and authoritarian styles are more common among large families, lower-income families, and families with younger adolescents.

6. Delinquency, sexual abuse, and suicide are among the most serious problems of adolescence.

About 80 percent of all adolescents in North America are guilty of minor law-breaking. Males are more likely to be arrested than females.

Our culture may promote minor law-breaking in boys as a way of establishing independence and asserting masculinity.

Delinquent young people have lower self-esteem, poorer family relationships, and more difficulty in school than their law-abiding peers.

Successful programs aimed at preventing delinquency are focused on developing skills in two specific areas: **school** and **employment.**

Relatives and family friends are the perpetrators in more than 75 percent of all cases of sexual abuse.

In **incestuous families,** the father is often introverted, immature, alcoholic, and has little social contact outside his family.

Adolescent victims of abuse tend to become involved again in violent relationships, either as the abuser or the abused.

Prevention involves desexualizing children in the mass media, encouraging fathers to participate in child-rearing, and promoting discussion of appropriate adult–child relationships as part of sex education.

Approximately 1 adolescent in every 10,000 commits suicide each year, a rate double that of twenty years ago.

Suicidal adolescents tend to be more solitary than normal adolescents, and show a greater tendency to be depressed, self-punishing, and emotional.

The warning signs of suicide include a sudden decline in school attendance and achievement, a break in a love relationship, withdrawal from social relationships, an attempted suicide, and cluster suicides.

Professional help for suicidal adolescents and their families focuses on opening channels of communication and keeping parents' expectations of their adolescent children in line with reality.

Answers to Testing Yourself

1. **c.** As many as 80% of all adolescents have broken the law in minor ways. (textbook, p. 377)

2. **b.** Males are much more likely to be arrested than females. (textbook, p. 377)

3. **a.** After rising in early adolescence, peer pressure to conform decreases throughout the remainder of this season. (textbook, p. 370)

4. **b.** A major theme of the audio program is that the image of the troubled adolescent as irretrievable is incorrect. (audio program)

5. **c.** Peer group pressure peaks at about age 14. (textbook, p. 370)

6. **c.** According to Erikson, forging an identity is the primary task of adolescence. (textbook, p. 363)

7. **d.** In doing so, the adolescent never forges a truly unique identity. (textbook, p. 364)

8. **a.** Having few commitments to goals or values, young people experiencing identity diffusion are often very apathetic. (textbook, p. 364)

9. **a.** Rites of passage are ceremonies or events—such as obtaining a driver's license—that provide a transition from one social status to another. (textbook, p. 368)

10. **d.** The typical adolescent friendship circle is quite large and fluid. (textbook, p. 372)

11. **d.** Peers perform a valuable role in helping the young person to develop an identity and reach maturity. (textbook, p. 373)

12. **c.** Democratic and equalitarian are the two styles recommended as healthiest by most psychologists. (textbook, p. 375)

13. **d.** Each of these family characteristics has been associated with autocratic and authoritarian parenting patterns. (textbook, p. 375)

14. **c.** Academic difficulty is an important predictor of delinquent behavior. (textbook, p. 377)

15. **d.** Just the opposite is true. Withdrawal from social relationships is a warning sign of suicide. (textbook, p. 385)

References

Jessor, R., & Jessor, S. (1977). *Problem behavior and psychosocial development: A longitudinal study of youth.* New York: Academic Press.

Professor Jessor, who is heard on the audio program, examines issues in high-risk adolescent behaviors.

Early Adulthood: Physical Development

AUDIO PROGRAM: Seasons of Eros

ORIENTATION

This is the first of a three-lesson unit on development between the ages of 20 and 40, the period of early adulthood. Lesson 17 focuses on physical development during this season.

In terms of our overall health, these years are the prime of life. Although physical decline progresses at the rate of about one percent per year, most changes go unnoticed because of **organ and muscle reserve.** Two exceptions noted in Chapter 17 of *The Developing Person Through the Life Span, 2/e,* are athletic performance and sexual response.

Physical development during early adulthood is not without potential problems, however. **Drug abuse, eating disorders,** and **violent death** are more likely during this season than at any other time. Chapter 17 discusses the possible causes of these problems and various means by which they may be prevented.

Audio program 17, "Seasons of Eros," explores how the expressions and meanings of sexuality, or **eros,** change over the life span. In a round table discussion featuring psychologists Janice Gibson, David Gutmann, and June Reinisch, psychiatrists Robert Butler and Thomas Carli, and psychotherapist Laura Nitzberg, it becomes clear that eros is more than intercourse; it is a capacity for pleasure that pervades the entire body and reflects the particular developmental needs of each individual and each age. The emerging life-span perspective makes it clear that, although its expressions and meanings may change, the life-giving force of eros manifests itself throughout the seasons of life.

LESSON GOALS

By the end of this lesson you should be prepared to:

1. Describe the normal age-related changes in physical growth, strength, health, and the sexual–reproductive system that occur during early adulthood.

2. Describe three major behavioral problems of young adulthood and discuss their possible causes and prevention.

3. Discuss the changing meaning of sexuality throughout the seasons of life.

Audio Assignment

Listen to the audio tape that accompanies Lesson 17: "Seasons of Eros."

Write answers to the following questions. You may replay portions of the program if you need to refresh your memory. Answer guidelines may be found in the Lesson Guidelines section at the end of this chapter.

1. Define "eros" and tell how its meaning changes from the beginning of life through adolescence.

At puberty – comes sex drive

2. Discuss the concept of sexual orientation and how experts believe it emerges in the individual.

3. Describe the ways in which the meaning and expression of eros change during adulthood.

Textbook Assignment

Read Chapter 17, "Early Adulthood: Physical Development," pages 391–409 in *The Developing Person Through the Life Span, 2/e.*

Write your answers to the following questions. Refer back to the textbook, if necessary. Answer guidelines may be found in the Lesson Guidelines section at the end of this chapter.

1. Describe the normal physical changes in growth, strength, and health that occur during early adulthood.

2. Identify the early physical signs of aging and explain why most signs of aging go unnoticed for many years.

3. Identify the age-related changes in sexual response that occur during early adulthood.

4. Discuss the problem of drug abuse in young adults, including its possible causes and prevention.

5. Discuss the dangers, causes, and prevention of eating disorders and destructive dieting.

6. Discuss the problem of violent death in young adults, including possible reasons for its prevalence and means of prevention.

Testing Yourself

After you have completed the audio and text review questions, see how well you do on the following quiz. Correct answers, with text and audio references, may be found at the end of this chapter.

1. "Eros" is most broadly defined as:
   a. the biological drive to reproduce.
   b. the drive to obtain pleasure and avoid pain.
   c. our species' natural attraction to members of the opposite sex.
   d. the desire for sexual and sensual pleasure.

2. Approximately _____ percent of the population is estimated to have a homosexual orientation.

   a. 1

   b. 5

   c. 10

   d. 20

3. Which of the following is true regarding the development of a homosexual orientation?

   a. Homosexuality is the product of a combination of biological and environmental influences.
   b. Homosexuality arises out of fear of the opposite sex.
   c. Most homosexuals were sexually molested as children.
   d. Homosexuality develops most readily in families with domineering mothers and weak, ineffectual fathers.

4. During which season(s) of life does eros tend to be diffused throughout the body and sense organs?

   a. Adolescence
   b. Early adulthood
   c. Later adulthood
   d. Both a and b

5. During which season(s) of life does eros tend to be most narrowly focused?

   a. Childhood
   b. Early adulthood
   c. Later adulthood
   d. Both b and c

6. One difference between men and women during early adulthood is that women have:

   a. a lower percentage of body fat.
   b. a greater conduction velocity of nerve fiber.
   c. a lower metabolism.
   d. greater organ reserve.

7. The first sign of aging likely to be noticed by adults in their late 20s is:

   a. a decline in physical endurance.
   b. a loss of hair.
   c. a decline in muscular strength.
   d. a decline in skin elasticity, resulting in wrinkles.

8. During early adulthood, the overall efficiency of most systems of the body:

   a. increases until about age 30.
   b. remains stable.
   c. gradually declines.
   d. declines more rapidly in men than in women.

9. Which of the following is an example of homeostasis?
   a. Shivering when we are cold
   b. Declining organ reserve as we get older
   c. Needing more rest after exercise when we are older
   d. the normal age-related decline in metabolism

10. As men grow older, they tend to:
    a. need less stimulation to become sexually aroused.
    b. experience a briefer interval between the beginning of sexual arousal and full erection.
    c. experience a longer refractory period.
    d. experience all of the above.

11. As women mature from early adolescence to middle adulthood, their sexual responsiveness tends to:
    a. increase.
    b. decrease.
    c. remain stable.
    d. become less predictable.

12. The condition in which fragments of the uterine lining become implanted on the surface of the ovaries or Fallopian tubes and block the reproductive tract is called:
    a. pelvic inflammatory disease (PID).
    b. in vitro fertilization.
    c. endometriosis.
    d. anovulation.

13. Each of the following problems is more prevalent during early adulthood than any other season *except*:
    a. drug abuse.
    b. compulsive eating.
    c. violent death.
    d. chronic disease.

14. Eating disorders such as bulimia and anorexia nervosa are often associated with:
    a. a low need for achievement.
    b. drug abuse.
    c. low stress tolerance.
    d. depression.

15. Compared to other men, the sons of alcoholic fathers are:
    a. no more likely to become alcoholic.
    b. much more likely to become alcoholic.
    c. much less likely to become alcoholic.
    d. more likely to abuse a variety of drugs.

LESSON 17 EXERCISE: EROS IN THE MEDIA

When sexuality is broadly defined as eros, it becomes clear that every season of life is sexual. Audio program 17 outlines the changing meanings of eros over the seasons of life.

Sexuality at various ages is often depicted in the popular media of television, magazines, motion pictures, novels, and radio. Television programs and advertising, for example, often portray sexual themes and stereotypes targeted to certain age groups.

The exercise for Lesson 17 asks you to look in the popular media for examples of advertising or programming that include portrayals of eros in various seasons. You may choose examples from current media sources or from your recollections of media portrayals in earlier seasons of your own life. After you have answered the following questions, return your completed **Exercise Response Sheet** to your instructor.

LESSON GUIDELINES

Audio Questions

1. When sexuality is broadly defined as **eros,** it becomes clear that every season of life is sexual. Children derive pleasure from touching their bodies, and from the sensory experiences of their eyes, ears, and mouths. In this first season, pleasure is diffused throughout the body.

   As children grow older, they become interested in each other's bodies and curious about where babies come from.

   Around the age of 6, children come to understand that our society is uncomfortable about sexuality and so they become more private, entering what Freud called the period of **latency.**

   In adolescence, eros is directed at others and concentrated in the sexual organs. Sexuality centers on mating and reproduction, although it may not be focused on a particular partner or necessarily coupled with intimacy.

2. In addition to forging their personal identity, each individual develops a sexual orientation toward intimacy with members of the same or opposite sex. Approximately ten percent of the population is homosexual.

   Experts still do not agree on what causes a person to develop a homosexual or heterosexual orientation. Some suggest that a biological predisposition is involved. One point of agreement is that sexual orientation does *not* appear to be learned. Children who grow up with gay or lesbian parents are no more likely themselves to be gay or lesbian than children who grow up with heterosexual parents. In all probability, it is a combination of certain biological factors and environmental influences that lead to the development of sexual orientation.

3. During early adulthood, eros is coupled with intimacy. A relationship with one special person becomes the basis for marriage and the bearing of children.

   Many couples report that the years of middle adulthood are the most sensual and sexual of their lives, as partners rediscover each other after the children have left home.

   Sex may become less frequent as one gets older, but it tends to be savored more fully with pleasure diffused over all the senses, just as it was at the beginning of life.

Textbook Questions

1. The early 20s are the prime of life in terms of physical development.

   Full height is reached at about age 18 in females and 20 in males.

   Increases in weight and body fat continue into the 20s.

   Before middle age, the average man adds 15 pounds and the average woman adds 14 pounds.

Women typically have a higher percentage of body fat and a lower metabolic rate than men do.

Since more of their body mass is muscle, men are typically stronger than women. In both sexes, physical strength increases during the 20s, reaching a peak at about age 30.

Digestive, respiratory, circulatory, and reproductive systems function at optimum levels during early adulthood. Cancer is the leading cause of death due to disease in young adults.

2. The first noticeable signs of aging (wrinkles, graying, or thinning of the hair) occur during the 20s.

   Beginning in the 20s, the efficiency of most body functions declines at a rate of about 1 percent a year.

   Life style, especially exercise, influences the rate of physical decline.

   The older a person is, the longer it takes for **homeostatic** adjustments, such as the regulation of body temperature, breathing, and heart rate, to occur.

   The physical decline of aging primarily affects **organ and muscle reserve**—the extra capacity that organs and muscles have to respond to unusual demands.

   With the exception of changes in **athletic performance** and the **sexual-reproductive system,** most of the age-related biological changes that occur in early adulthood are of little consequence.

3. In most men, age-related declines in sexual responsiveness are usually not a concern until middle or late adulthood.

   As men grow older, they often need more direct or explicit stimulation to produce sexual arousal.

   In older men, a longer time elapses between the beginning of sexual excitement and full erection, between erection and ejaculation, and between orgasm and the end of the refractory period.

   Age-related trends in sexual responsiveness are not as clear for women.

   In general, women become more likely to experience orgasm as they mature from early adolescence to middle adulthood. This may be a result of the more prolonged stimulation provided as a consequence of the slowing of men's responsiveness.

   Approximately 15 percent of all couples have fertility problems.

   In males, the most common fertility problem is a low number of sperm and/or poor sperm **motility.**

   Most women find that ovulation becomes less regular as middle age approaches. It therefore takes longer to conceive, and they are more likely to have twins when they do.

   Another common fertility problem in women, often caused by PID (pelvic inflammatory disease), is blocked Fallopian tubes.

   Endometriosis is most common in women between the ages of 25 and 35 and accounts for approximately 15 percent of fertility problems.

Many fertility problems, such as low sperm count or blocked Fallopian tubes, can be solved with modern medical techniques.

Another alternative is **in vitro fertilization** (outside the uterus), which today has a success rate of about 10 percent.

Most doctors recommend that women begin their child-bearing before age 35.

4. **Drug abuse** is defined as use of a drug that impairs an individual's physical, cognitive, or social well-being.

   Genetic predisposition, sex, temperament, and family and cultural context are major factors in determining an individual's propensity to use and respond to a given drug.

   Young adulthood is the time when problem drinking and drug abuse are the most likely, and the most likely to result in damage. Four possible reasons are discussed: (a) many young adults abuse drugs as a way of striving for independence; (b) several life stresses, such as finishing school, beginning a family, and establishing a career, are clustered during early adulthood; (c) the need to feel sexually attractive is very intense during early adulthood and many young adults feel that alcohol and other drugs enhance sexual responsiveness; (d) the life style of many young adults promotes alcohol and drug use.

   The use of legal and illegal drugs peaks at about age 23. Seventy-five to 80 percent of today's young adults have tried an illicit drug (most often marijuana or cocaine).

   Drug abusers often miss work or school, neglect or abuse their sexual partner, commit crime to support their habit, or become involved in acts of violence. They are also involved in most fatal single-car crashes, homicides, and many suicides.

   The ability to master the developmental tasks of young adulthood is impaired by the irrationality and social misjudgment that drug abuse causes.

   Measures for preventing drug abuse in early adulthood include education based on accurate information, prevention tailored to the individual, behavior modification, and stiff penalties for drunk driving.

5. There is no evidence that being a little fat is a health hazard.

   Cultural standards of beauty, however, are based on a low percentage of body fat. Such restrictive stereotypes about beauty probably promote the eating disorders that afflict many women.

   Continual worry about weight and repeated dieting can be destructive:

   Too low a percentage of body fat is associated with irregularities in the menstrual cycle.

   Crash diets often produce nutritional imbalance.

   Extensive dieting lowers the metabolic rate and may actually promote weight gain.

Diet failures often produce feelings of frustration, guilt, and depression.

Addiction to diet pills may result from some weight-loss regimens.

Bulimia and **anorexia nervosa** may be life-threatening.

Young college women are especially likely to diet, take diet pills, binge, and purge.

6. Stereotypes about "masculine" behavior may promote violent death due to accident, suicide, or homicide, particularly in young adult men.

One American male in every forty between the ages of 15 and 35 dies a violent death. This rate is four times the incidence of violent death in women of the same age.

Living up to the ideals of masculinity makes it hard for young men to avoid conflicts, back away from challenges, or admit that they need help.

The prevalence of handguns (and lack of restrictions on their use) is a significant factor in the high rate of violent death in the United States.

Homicide is the leading cause of death for young adult black men, which suggests that the negative social forces of urban ghettos (high rates of crime, drug use, single-parent families, unemployment) are factors contributing to the high incidence of violent death.

Accident rates vary with ethnicity and the incidence of alcoholism. Compared to other ethnic groups in the United States, Native Americans have the highest rate of alcoholism (related to a genetic predisposition to addiction) and twice the rate of fatal accidents.

Compared to the research efforts expended to prevent the leading causes of death of middle-aged and older adults, very little has yet been devoted to the prevention of violent death in young adults.

Answers to Testing Yourself

1. **d.** Eros is defined as the desire for sexual and sensual pleasure. (audio program)

2. **c.** About 10 percent of the population will eventually identify itself as homosexual. (audio program)

3. **a.** Homosexuality is most likely due to a combination of biological and environmental influences. (audio program)

4. **c.** During later adulthood, as in childhood, eros is spread throughout the body and senses. (audio program)

5. **b.** During early adulthood, eros tends to be focused on reproduction. (audio program)

6. **c.** Women typically have a lower metabolism than men do. (textbook, p. 392)

7. **d.** Most people notice the first signs of aging in their physical appearance. (textbook, p. 392)

8. **c.** The efficiency of most body functions begins to decline in the 20s. (textbook, p. 393)

9. **a.** Shivering serves to activate the muscles and warm the body—an example of homeostatic adjustment. (textbook, p. 393)

10. **c.** As men grow older, a longer time elapses between orgasm and the end of the refractory period. (textbook, p. 397)

11. **a.** In general, as women mature from early adolescence toward middle adulthood, they become more likely to experience orgasm. (textbook, p. 397)

12. **c.** Endometriosis is the likely cause for about 15 percent of fertility problems. (textbook, p. 399)

13. **d.** In fact, during early adulthood our bodies are healthier than during any other period. (textbook, p. 391)

14. **d.** Eating disorders often go hand in hand with depression. (textbook, p. 406)

15. **b.** Sons of alcoholic fathers are four times as likely to become alcoholics themselves as are sons of nonalcoholics. (textbook, p. 400)

References

Daly, Martin, & Wilson, Margo. (1978). *Sex, evolution & behavior*. Belmont, California: Wadsworth.

An interdisciplinary work on sexuality that includes discussion of human sexuality, sex role development, and parental care.

Whitbourne, Susan Krauss. (1985). *The aging body*. New York: Springer-Verlag.

A comprehensive description of age-related changes in physical vitality.

Early Adulthood: Cognitive Development

AUDIO PROGRAM: The Development of Faith

ORIENTATION

Lesson 18 is the fifth in a sequence of lessons that track cognitive development from infancy to late adulthood. The first two of these, Lessons 6 and 9, explored the acquisition of language. The next two, Lessons 12 and 15, presented the theory of Jean Piaget, who identified four stages of cognitive development culminating in the abstract logic of formal operations. In this lesson we step beyond Piaget and explore other ways of thinking.

As people grow from adolescence into adulthood, the commitments, demands, and responsibilities of adult life produce a new type of **postformal thinking** that is better suited than formal operations to solving the practical problems of daily life. Postformal thought is more adaptive, flexible, and **dialectical.** Dialectical thought, which some researchers consider the most advanced form of cognition, recognizes that most of life's important questions do not have single, unvarying, correct answers. It is grounded in the ability to consider both sides of an issue simultaneously. Chapter 18 of *The Developing Person Through the Life Span, 2/e,* contrasts postformal with formal operational thought and describes the stages of adult cognitive development proposed by K. Warner Schaie. Schaie's stages correspond to the patterns of commitment to personal goals, family, and society that are generally accepted as typical of adult life.

Thinking about questions of **faith** and ethics may also progress during adulthood, especially in response to significant life experiences such as participating in higher education and becoming a parent. Audio program 18, "The Development of Faith," presents James Fowler's theory of how faith changes throughout the seasons of life. Fowler, a Christian minister and professor of theology, has bridged the fields of psychology and religion by creating a model in which faith is broadly conceived. Building on the cognitive and personality theories of Jean Piaget and Erik Erikson, Fowler's model extends the concept of faith beyond religious faith to include whatever each person really cares about—his or her "ultimate concern." Fowler's theory describes six stages, each of which has distinct features and is classified as typical of a certain age.

Although Fowler's theory is not without its critics, its emphasis on faith as a developmental process rings true. If Fowler is correct, faith, like other aspects of cognition, may mature from the simple self-centered and one-sided perspective of the child to the much more complex, altruistic, and multifaceted perspective of many adults.

LESSON GOALS

By the end of this lesson you should be prepared to:

1. Describe cognitive development during early adulthood and the stages of adult cognition proposed by K. Warner Schaie.

2. Discuss the complexity of postformal thinking and moral reasoning.

3. Describe the relationship of adult cognitive growth to higher education and other significant life events.

4. Outline James Fowler's stage theory of the development of faith.

Audio Assignment

Listen to the audio tape that accompanies Lesson 18: "The Development of Faith."

Write answers to the following questions. You may replay portions of the program if you need to refresh your memory. Answer guidelines may be found in the Lesson Guidelines section at the end of this chapter.

1. List and describe the six stages in the development of faith proposed by James Fowler.

2. Discuss the relationship between cognitive development and faith.

3. Cite several criticisms of Fowler's theory.

Textbook Assignment

Read Chapter 15, "Early Adulthood: Cognitive Development," pages 411–429 in *The Developing Person Through the Life Span, 2/e.*

Write your answers to the following questions. Refer back to the textbook, if necessary. Answer guidelines may be found in the Lesson Guidelines section at the end of this chapter.

1. Describe three approaches to the study of adult cognition.

2. Outline the five stages in cognition proposed by K. Warner Schaie.

3. Contrast formal operational thought with postformal thought.

4. Define dialectical reasoning and discuss the relationship between cognitive growth and moral reasoning.

5. Discuss the relationship between cognitive growth and higher education.

6. Cite several ways in which life events may foster cognitive development.

Testing Yourself

After you have completed the audio and text review questions, see how well you do on the following quiz. Correct answers, with text and audio references, may be found at the end of this chapter.

1. According to James Fowler, the simplest stage of faith is the stage of:
   a. universalizing faith.
   b. intuitive-projective faith.
   c. mythic-literal faith.
   d. conventional faith.

2. At one stage in the development of faith, people learn to question the practices and philosophies of significant persons in their lives. This is the stage of:
   a. conjunctive faith.
   b. individual-reflective faith.
   c. synthetic-conventional faith.
   d. intuitive-projective faith.

3. At the highest stages of faith development, people have incorporated a powerful vision of compassion and human brotherhood into their lives. This stage is called:
   a. conjunctive faith.
   b. individual-reflective faith.
   c. synthetic-conventional faith.
   d. universalizing faith.

4. The stage of faith that corresponds with the age at which most individuals achieve concrete operational thought is called:
   a. mythic-literal faith.
   b. individual-reflective faith.
   c. synthetic-conventional faith.
   d. universalizing faith.

5. For which of the following has Fowler's theory of faith been criticized?
   a. For focusing primarily on religious faith
   b. For proposing that faith develops in stages, rather than continuously
   c. For proposing stages of faith that many people will never reach
   d. For all of the above reasons

6. According to K. Warner Schaie, during early adulthood, adult cognition is in the:
   a. acquisition stage.
   b. achieving stage.
   c. responsible stage.
   d. reintegrative stage.

7. Most of the problems of adult life are characterized by ambiguity, partial truths, and an infinite number of variables that occur in:
   a. open systems.
   b. closed systems.
   c. concrete operational systems.
   d. formal operational systems.

8. The ability to consider both poles of an idea simultaneously is the hallmark of:
   a. concrete operational thought.
   b. formal operational thought.
   c. dialectical thought.
   d. conjunctive thought.

9. According to the textbook, whether a person achieves postformal thought depends primarily on:
   a. his or her education.
   b. his or her life experiences and education.
   c. his or her age.
   d. his or her cultural background.

10. In comparison to college students of a generation ago, today's students are more likely to:
    a. choose a liberal-arts curriculum.
    b. postpone college for a year or two.
    c. choose a career-based curriculum.
    d. value nonacademic extracurricular activities.

11. According to the textbook, an especially important factor in determining whether college students learn to think deeply is:
    a. the availability of faculty members as role models and mentors.
    b. a low faculty to student ratio.
    c. the general intelligence of the student body.
    d. all of the above.

12. In K. Warner Schaie's theory, middle-aged adults often temper their need for personal achievement with a greater concern for their family. This is called the:
    a. reintegrative stage.
    b. achieving stage.
    c. executive stage.
    d. responsible stage.

13. According to the textbook, a typical effect of college education is that it leads students to become:
    a. very liberal politically.
    b. less committed to any particular ideology.
    c. less tolerant of others' views.
    d. more tolerant of others' views.

14. A central idea of Schaie's theory is that prior to adulthood, cognitive development is:
    a. unrelated to the specifics of an individual's life.
    b. closely related to individual experiences.
    c. especially diverse.
    d. largely the product of the school system.

15. Which of the following is *not* one of the major approaches to the study of adult cognition described in the textbook?
    a. The information-processing approach
    b. The developmental approach
    c. The systems approach
    d. The psychometric approach

LESSON 18 EXERCISE: THINKING DURING ADULTHOOD

One theme of the *Seasons of Life* series is that every age brings a different way of knowing. The cognitive patterns that emerge during adulthood are propelled by the commitments each individual makes during this time. These include commitments to personal achievement, family concerns, and the community at large. Such commitments give the individual a new understanding of the complexity of most of life's daily problems.

In this lesson we have explored several types of thinking, including formal and postformal thought, dialectical reasoning, and faith. Test your understanding of these ways of thinking by writing answers to the following questions. Return your completed **Exercise Response Sheet** to your instructor.

LESSON GUIDELINES

Audio Question Guidelines

1. Stage One: **Intuitive-projective faith** refers to the imaginative faith that emerges as children acquire the use of symbols and language.

   Stage Two: **Mythic-literal faith** corresponds to Piaget's stage of concrete operations. In this stage children become interested in learning the stories of their culture, and often take these stories literally.

   Stage Three: **Synthetic-conventional faith** often emerges during adolescence as a result of the young people's new awareness of who they are and what they believe in. Faith is characterized by a conformist, nonintellectual acceptance of the values and ideals of people who are important to the young person.

   Stage Four: **Individual-reflective faith** often begins in early adulthood, as individuals become critically reflective of their beliefs. Characterized by intellectual detachment from the values of their culture and the approval of others, faith in this stage may represent God in the abstract or as a philosophical concept. In the audio program, Fowler refers to this stage as "individuative."

   Stage Five: **Conjunctive faith** rarely develops before middle age. Considered the highest stage that most people experience, conjunctive faith recognizes the many paradoxes and inconsistencies of life.

   Stage Six: **Universalizing faith** refers to the behavior of rare individuals who develop a vision of universal compassion, justice, and love that often leads to the denial of their personal welfare in an effort to serve these beliefs.

2. James Fowler has delineated six stages of faith that progress from a simple, self-centered perspective to a more complex, altruistic, and multifaceted view.

   In developing this theory, Fowler was strongly influenced by Piaget's stages of cognitive development. Although Fowler's stages are not considered exclusive to a given age range, each is classified as typical of a certain age group and a certain stage of cognitive development. In the first stage of faith (intuitive-projective faith), for example, egocentric preschoolers, who are unable to take the perspective of another person, often form a highly imaginative and nonhuman image of God. In the second stage (mythic-literal faith)—equivalent to Piaget's stage of concrete operations—children begin to understand cause-and-effect relationships, but are limited to the concrete reality of the here and now. Their faith puts a correspondingly literal interpretation on the stories and myths of their religion and culture.

   Later in adulthood, the development of faith is also paced by cognitive development. When postformal thought is achieved, for example, faith can become dialectical in nature and recognize the often paradoxical nature of life. This is what happens in the conjunctive stage. At each season of life, therefore, the development of faith and thinking go hand in hand.

3. Like any pioneer, James Fowler is not without his critics. One criticism is that although Fowler defines faith broadly, he overemphasizes religious faith.

Another criticism is that faith may develop continuously, rather than in stages as Fowler proposed.

A third objection is based on the reluctance of some critics to accept the notion that some stages of faith are higher than others.

Textbook Question Guidelines

1. The **developmental approach** emphasizes cognitive stages and processes that change with age.

   The **psychometric approach** focuses on changes in the components of intelligence as measured by IQ tests.

   The **information-processing approach** is primarily concerned with how the efficiency of mental processes, such as input, encoding, memory, and output, are affected by age.

   The textbook takes a developmental approach to the study of cognition by describing age-related changes in cognition in an effort to discern patterns or predictable stages.

2. Compared to adolescent thinking, adult thinking is less self-centered, broader, more reasonable, and more practical. These changes come in response to the individual assuming the responsibilities of adult life.

   K. Warner Schaie has proposed four stages of adult cognition that correspond to the patterns of commitment of adult life.

   During the **period of acquisition,** information is absorbed indiscriminately, with little regard for its practical significance.

   During late adolescence young people enter the **achieving stage,** during which learning is more goal-directed.

   As middle adulthood approaches, many people enter the **responsible stage,** during which the entrepreneurial style of the achieving stage is replaced by a pattern that integrates personal goals with family goals.

   Some adults develop an unusually broad sense of social responsibility and enter what Schaie calls the **executive stage.**

   During late adulthood, the **reintegrative stage** is achieved. Adults turn to making sense out of their life as a whole.

3. Formal operational thought permits the individual to engage in hypothetical-deductive reasoning and to coordinate a variety of logical relationships. Formal thinking is best suited to **closed systems** in which there are a limited number of known variables.

   Most daily problems occur in **open systems**—family, work, friends, community—that are ambiguous and contain an infinite number of variables.

Postformal thought is less abstract, less intellectual, and less absolute than formal operations. It is also dialectical, adaptive, and more complex because it takes emotions, motives, and personal experiences into consideration.

4. **Dialectical thinking** involves considering both sides of an idea at the same time, and recognizing that most of life's important questions do not have single, unchangeable, correct answers.

   Many researchers, including Carol Gilligan and Lawrence Kohlberg, believe that the responsibilities of adulthood also promote higher moral reasoning.

   Gilligan believes that as life experiences expand, people often shift from ideological or personal moral reasoning to moral reasoning that emphasizes ethical responsibility.

5. Level of education is highly correlated with many measures of adult cognition.

   College education leads people to become more tolerant of differing views, more flexible and realistic in their attitudes, and dialectical in their reasoning.

   Based on an intensive study of Harvard students, William Perry found that the thinking of college students progressed through nine levels of complexity, becoming more relativistic and accepting of multiple views.

   The impact that any college has on an individual's cognitive growth depends on many factors, including the availability of professors as mentors and role models.

6. It has been suggested that specific life events often foster cognitive development. Becoming a parent, for example, may offer new insights, leading to a more responsible and less self-centered view of the world.

   Other life events that might promote cognitive development include being promoted or fired from a job, exposure to different lifestyles, religious experiences, and the death of a loved one.

Answers to Testing Yourself

1. **b.** In this stage faith is magical, illogical, imaginative, and filled with fantasy. (audio program; textbook, p. 418)

2. **b.** A person's ability to articulate his or her own values, separately from family and friends, is characteristic of individual-reflective faith. In the audio program, Fowler calls this stage "individuative." (audio program; textbook, p. 418)

3. **d.** Persons reaching stage six of faith (examples include Mahatma Gandhi, Martin Luther King, Jr., and Mother Teresa) are exceedingly rare. (audio program; textbook, p. 419)

4. **a.** At this stage, the individual takes the myths and stories of his or her religion literally. (audio program; textbook, p. 419)

5. **d.** Each of these has been offered as a criticism of Fowler's theory. (audio program)

6. **b.** The thinking of young adults in this stage is very goal-directed. (textbook, p. 413)

7. **a.** Thinking that is required in open systems is best accomplished with postformal, rather than formal, operational thought. (textbook, p. 414)

8. **c.** Some researchers consider dialectical thought to be the most advanced form of thinking. (textbook, p. 415)

9. **b.** Postformal thought is usually not attained until a combination of adolescent maturation and educational experience makes it possible. (textbook, p. 414)

10. **c.** Some researchers believe that the prevalence of the liberal arts curriculum in colleges twenty years ago accounted for the clear development in dialectical thinking observed among college students. (textbook, p. 422)

11. **a.** The proper match of students with teachers is identified as an especially important variable in determining the influence of college education on cognitive development. (textbook, p. 424)

12. **d.** While the younger person might have a clear goal of personal achievement, in ten years or so he or she typically adjusts this goal to fit in with family goals. (textbook, p. 413)

13. **d.** Longitudinal research has consistently shown that college education leads people to become more tolerant of political, social, and religious views that differ from their own. (textbook, p. 421)

14. **a.** Schaie begins with the distinction that prior to early adulthood, cognitive development is largely unrelated to the specifics of the individual's life; afterwards, it is closely tied to them. (textbook, p. 412)

15. **c.** Each of the other three perspectives is identified in the textbook as valid and useful to the study of cognitive development. (textbook, p. 411)

References

Fowler, James W. (1981). *Stages of faith: The psychology of human development and the quest for meaning.* New York: Harper and Row.

James Fowler outlines his influential theory of the development of faith.

Early Adulthood: Psychosocial Development

AUDIO PROGRAM: Not Being First

ORIENTATION

Lesson 19 of *Seasons of Life* is concerned with psychosocial development in early adulthood and the changing composition of the American family. Chapter 19 of the textbook discusses the developmental courses of love, marriage, career, and parenthood, along with the special problems of divorce, remarriage, and stepfamilies.

For many people, the question "Who's in your family?" is difficult to answer. Although many people tend to form very close-knit nuclear families consisting of a mother, a father, and one or more children, the number of **stepfamilies** is increasing. This increase is not due, as it was in the past, to death and remarriage, but to divorce and remarriage. 1974 was the first year more marriages in the United States were ended by divorce than by death. Since people are living longer now than at any other time in history, divorced persons have more opportunities to remarry.

These trends in the composition of American families have created new notions about the words "family," "mother," and "father." Unlike nuclear families, stepfamilies usually include children who are members of two households. Extra sets of in-laws and grandparents and the stress of the competition that often exists between a stepparent and the ex-spouse all serve to complicate family relationships.

The audio program "Not Being First" poignantly illustrates the particular dilemma of the stepparent by introducing the listener to Penny and Lyn Beesley. Married for six years, both Penny and Lyn have had previous marriages that ended in divorce. As the program unfolds, we hear of Lyn's struggles to form a bond with Heather, Penny's daughter from her first marriage. We also hear commentary from counselor Elaine Horigian, a clinical psychologist, and Helen Weingarten, a professor of social work.

LESSON GOALS

By the end of this lesson, you should be prepared to:

1. Discuss the ways in which adults meet their needs for love/affiliation and work/achievement during early adulthood.

2. Describe how the structure of the typical American family has changed, the causes of these changes, and the problems faced by stepparents and stepchildren as a result of more complicated family structures.

3. Discuss the issues facing adults in selecting a mate, marrying, choosing a career, and parenting.

Audio Assignment

Listen to the audio tape that accompanies Lesson 19: "Not Being First."
    Write answers to the following questions. You may replay portions of the program if you need to refresh your memory. Answer guidelines may be found in the Lesson Guidelines section at the end of this chapter.

1. In what ways do **nuclear families** and **stepfamilies** differ?

2. How has the prevalance of stepfamilies changed from the seventeenth century to the present day? What are the major causes of this change? Do experts predict that this trend will continue?

3. What are the three stages in stepfamily development described by clinical psychologist Elaine Horigian?

4. What are some of the typical problems faced by the members of stepfamilies?

5. Why is it that "Every person in a stepfamily has experienced a significant loss?"

6. What are some of the long-range effects on children of living in a stepfamily?

Textbook Assignment

Read Chapter 19: "Early Adulthood: Psychosocial Development," pages 431–458 in *The Developing Person Through the Life Span, 2/e.*

Write your answers to the following questions. Refer back to the textbook, if necessary. Answer guidelines may be found in the Lesson Guidelines section at the end of the chapter.

1. What are the basic psychosocial needs of early adulthood according to Maslow, Erikson, Freud, and other developmental psychologists?

2. What is the typical path of career development in early adulthood?

3. What are the stages in the process of mate selection?

4. What are the stages of parenthood in the life cycle?

Testing Yourself

> After you have completed the audio and text review questions, see how well you do on the following quiz. Correct answers, with text and audio references, may be found at the end of this chapter.

1. The most basic psychosocial needs of early adulthood are:
   a. love and belonging.
   b. affiliation and achievement.
   c. generativity and stagnation.
   d. passionate and companionate love.

2. For most adults, the late 20s are characterized mainly by:
   a. stabilization and accommodation.
   b. rebellion against society.
   c. one final fling before settling down.
   d. exploring alternative lifestyles.

3. What is the primary function of the adult "social clock"?
   a. To record important social events in one's life
   b. To tell each person when to eat, fall asleep, etc.
   c. To help determine if behavior is developmentally early, on time, or late
   d. To serve as a daily reminder of events on one's social calendar

4. The first stage of career development is:
   a. career exploration.
   b. the advancement period.
   c. the establishment phase.
   d. research and development.

5. The parents of a two-year-old child are in the _____ stage of parenthood.
   a. authority
   b. nurturing
   c. interpretive
   d. interdependent

6. Which of the following best describes a mentor?
   a. A loving parent who helps a child
   b. A good friend who helps a friend
   c. An older individual who helps a younger one
   d. A technical consultant who helps clients

7. What does it mean when we say that in recent years the social clock has relaxed its norms for American women?
   a. Most important life events are taking place at an older age.
   b. Most important life events are taking place at a younger age.
   c. With longer lives and fewer children, women are present in the classroom and the work place at all ages.
   d. Women have finally achieved equity with men in the work place.

8. In terms of marriage, what is meant by homogamy?

   a. Marriage between one woman and one man
   b. Marriage between a man and a woman who are similar in socioeconomic status, age, religion, and other ways
   c. Marriage between a man and a woman who are dissimilar in socioeconomic status, age, religion
   d. Marriage between two people of the same sex

9. Married adults are now divorcing at twice the rate that adults did twenty years ago. Which of the following has *not* been suggested as a factor in the increased divorce rate?

   a. Divorce laws have made it easier to obtain a divorce today.
   b. Alternatives to marriage have become more acceptable.
   c. Today's spouses expect more understanding and communication from their mates.
   d. Adults today are less emotionally mature.

10. The divorce rate for remarriage is:

    a. dropping by approximately 25 percent each year.
    b. exactly the same as that of first marriages.
    c. slightly higher than that of first marriages.
    d. half that of first marriages.

11. In the 17th century most marriages were ended _____; today most are ended _____.

    a. by divorce . . . by the death of a spouse.
    b. by the death of a spouse . . . by divorce.
    c. before children were born . . . after children are born.
    d. after children were born . . . before children are born.

12. Compared to the 19th century, the number of stepfamilies today is:

    a. significantly greater.
    b. about the same.
    c. a little greater.
    d. significantly less.

13. According to experts, one common mistake that many stepfamilies make is to:

    a. try to become the same as nuclear families.
    b. allow their stepchildren to do anything they want.
    c. discipline their stepchildren too harshly.
    d. avoid establishing any "deep" relationships within the stepfamily.

14. Concerning children within stepfamilies, which of the following is true?

    a. Small children tend to form new loyalties only with difficulty.
    b. Small children tend to think in terms of absolutes such as, "You can only love one mom and one dad."
    c. Small children often feel resentful and hostile toward a stepparent.
    d. All of the above are true.

15. According to counselor Elaine Horigian, in the final stage of stepfamily living
    a. family members "let go" of unrealistic expectations and realize that their stepfamily will never be the same as a nuclear family.
    b. each family member "bends over backwards" trying to please other family members.
    c. family members become alienated to the extent that a sense of closeness is impossible.
    d. none of the above occurs; it is impossible to predict such stages.

LESSON 19 EXERCISE: GENOGRAMS

As an exercise in studying the trends described in the audio program, and reflecting on your own life experiences, construct a **genogram** of your own, or another, family. A genogram is a map of several generations within a family, something like a family tree. By convention, in genograms males are represented by squares and females by circles. Marriage is indicated by a solid line drawn from circle to square, and divorce by a dotted line. Death is indicated by drawing an "X" through the circle or square. The genogram is expanded horizontally to include additional individuals within a given generation, and vertically to document the family history across several generations. Here, for example, is the beginning of a genogram representing the Beesley family, who were introduced in the audio program.

Jack ——— Penny ——————————— Lyn ——— Shelly

Heather

This genogram would be completed by adding Penny's and Lyn's brothers, sisters, parents, and grandparents.

Since the turn of the century, family relationships and the typical structure of family genograms have changed in several ways. For one thing, families with multigenerational living members are much more common. For another, the average size of nuclear families has declined and there are more single-parent and stepparent households. These changes are resulting in more complex genograms. As Lyn Beesley said of Heather's family tree, "It had a whole lot of branches on it!" Is the same going to be true of your genogram?

LESSON GUIDELINES

Audio Question Guidelines

1. Stepfamilies are much more complex than nuclear families.

   Stepfamilies often result in children becoming members of two households.

   In stepfamilies, competition may exist within and among relationships.

   In stepfamilies there are more relationships: extra inlaws, grandparents, and so on.

2. In the 17th century, stepfamilies were very common due to the deaths of spouses and a high rate of remarriage among widows and widowers.

   In the 18th and 19th centuries, the number of stepfamilies declined.

   In the middle of the 20th century, as divorce rates began to soar and people continued to live longer than in the past, the number of stepfamilies began to rise.

   Although it is not easy to count the number of stepfamily households, the rising trend is expected to continue.

3. Stage One (The "Honeymoon" Stage): For approximately six months after the remarriage, everyone "tries too hard," and avoids dealing with their true feelings of fear, resentment, and threat.

   Stage Two (The "Conflict" Stage): Negative feelings start to come out. It's necessary for the natural parent to support the stepparent and give him or her credibility.

   Stage Three (The "Letting Go" Stage): The stepfamily begins to realize that it will never be the same as a nuclear family and lets go of some of the myths, ideals, and dreams it once held. This is a very rewarding stage because the stepfamily members realize that their relationships are deep and caring, even though they are different from those within a nuclear family.

4. The new stepfamily is often immediately faced with a set of very difficult problems involving deeply held feelings of loyalty.

   Stepfamilies often attempt to re-create the exclusive, close-knit relationships of a nuclear family.

   The extra relationships within stepfamilies may create feelings of threat and competition.

   It takes a long time for children to develop new loyalties. Stepparents often are disappointed that their relationships with stepchildren are not as close as they would like.

   Every person in a stepfamily has experienced some sense of loss.

   Stepparents may feel alienated, awkward, and as if they are constantly "walking on eggs."

   With so many diverse relationships within a stepfamily, social etiquette is often difficult. Graduations and weddings can pose problems, for example.

Issues of sexuality are different than they are in nuclear families. Sexual feelings between husband and wife, between stepparents and stepchildren, and between unrelated youngsters suddenly brought together must all be dealt with.

5. Children may feel that they have "lost" a natural parent.

Stepparents lose the sense of "being first" in their relationships with stepchildren.

Stepparents experience a sense of loss of the dreams that accompany a first-time marriage.

Parents feel a sense of loss of their former spouse.

6. Stepfamilies create environments in which children may be inundated with adults. The effects of this type of environment on character development are unknown.

With more children becoming members of two households, some experts are concerned that such children may have too easy a way out of their problems, with too many adults ready to give the answers children are seeking.

Textbook Question Guidelines

1. Although psychosocial development in early adulthood is marked by diversity, two basic needs drive its development: affiliation and achievement or, expressed in other terminology, **social acceptance** and **competence.** Other theorists describe these needs as follows:

Maslow **love and belonging,** and the need for **success and esteem.**

Freud: **love and work.**

Erikson: The crises of **intimacy vs. isolation** and **generativity vs. stagnation.**

Levinson believes that the years of adulthood follow an orderly sequence of life-structure building and life-structure changing. Periods of questioning and change, or **transitions,** occur in every decade.

2. For many adults, working is central to living for more than economic reasons.

During the first stage, **career exploration,** the person attempts to match his or her skills to a specific job. Young adults tend to be less satisfied with their jobs than are older workers. By age 30, the average American adult has held three distinct occupations.

Usually at about age 30 a worker enters the second phase of career establishment, in which he or she is given more responsibility, money, and respect. During this period workers tend to work harder and to be more satisfied with their work.

Key factors in job satisfaction are the opportunity to choose one's own hours of work (**flextime**), and a relationship with a more experienced worker, or **mentor.**

3. **Stimulus stage.** A stimulus such as appearance, intellect, popularity, or status makes two individuals notice each other. Physical appearance is a powerful stimulus, especially for men. Women are also drawn by status, and prefer men who are leaders, have more education, better jobs, and so on.

   Values stage. The couple compares their values, including attitudes toward work, marriage, religion, culture, and society. The greater the similarity in the couple's views, the more likely their attraction is to deepen.

   Roles stage. As the relationship deepens, the couple begins to develop roles with each other. The partners discover how each copes with stress, anger, and the situations of daily life. Men tend to think they are compatible with their partners sooner than do women, who are more cautious.

4. Based on the age of the oldest child, the stages of parenthood begin with the **honeymoon period,** during which the newly married couple establishes their relationship to each other and decides whether or not to become parents.

   During the **nurturing period** (birth to age 2) parents share the joys and demands of caring for a new infant. Many mothers have unrealistic expectations about this period, as they look forward to quitting their jobs and focusing time on their family and homes. Many new fathers feel that the arrival of the new baby results in more demands being placed on them, with reduced love and attention from their wives.

   The **authority period** (2–5 years) is often the period of greatest direct confrontation between husband and wife, due to increased financial burdens, the child's growing needs, shifting roles, and often the birth of a second child.

   During the **interpretive stage** (6–11 years) parenting becomes easier as children become more self-sufficient and enter school.

   During the **interdependent period** of adolescence, parenthood is more of a challenge, as the child demands the privileges of adulthood before being able to handle them responsibly.

Answers to Testing Yourself

1. **b.** Although various researchers use different terminology, the most basic needs of adulthood are affiliation and achievement. (textbook, pp. 431–432)

2. **a.** By their late 20s, most adults have completed the move from the period of late adolescent exploration to adult stabilization. (textbook, pp. 433–434)

3. **c.** The social clock is a culturally set timetable that establishes when various events in life are appropriate and called for. (audio program, textbook, p. 434)

4. **a.** In this first stage the person attempts to match his or her skills, talents, and needs to a specific occupation and job. (textbook, p. 451)

5. **a.** The authority period spans the preschool years, between ages 2 and 5. (textbook, pp. 452–453)

6. **c.** For many workers a key factor in job satisfaction is a relationship with an older, more experienced worker who fills many roles—teacher, protector, confidant, and friend. (textbook, p. 451)

7. **c.** The social clock varies historically. In contemporary American culture, the social clock now allows more diversity than it once did. (audio program, textbook p. 436)

8. **b.** Marriage between individuals who are similar in age, socioeconomic background, religion, and other factors, has a better chance of survival than does one based on heterogamy. (textbook, p. 439)

9. **d.** The easing of divorce laws, increasing acceptability of alternatives to staying married, and greater expectations of mates have all contributed to the rise in divorce. There is no evidence that emotional maturity or depth of commitment is any less today than in previous years. (textbook, pp. 444–446)

10. **c.** The divorce rate for remarriages is higher than that for first marriages, in some instances because lonely divorced people remarry too quickly. (textbook, p. 448)

11. **b.** Today, with higher divorce rates and longer life expectancy, more and more people are marrying for the second or third time. (audio program)

12. **a.** Today, approximately one child in four will spend some time living with a stepparent before turning 17. (audio program)

13. **a.** A stepparent can never replace a natural parent; to attempt to do so is, according to experts, disastrous. (audio program)

14. **d.** For all of these reasons, stepparents are often disappointed that they cannot quickly develop close relationships with their stepchildren. (audio program)

15. **a.** According to Horigian, this final stage of "letting go" is very rewarding. Once families are freed from the disappointment of lost dreams, the relationships within a stepfamily are released to grow in their own unique ways. (audio program)

References

Schneider, D.M. (1980). *American kinship: A cultural account.* Englewood Cliffs, NJ: Prentice-Hall.

An interesting anthropological/sociological description of cultural variations in family structure.

Middle Adulthood: Physical Development

ORIENTATION

How long do you think you will live? What are the odds that you will survive to be 100 years old? Is there anything you can do to improve these odds? An individual's **longevity** is limited by the biological clock, but the limit is flexible. As the author of *The Developing Person Through the Life Span* indicates, today's cohort of middle-aged persons is healthier than in previous years. Although many physical changes occur between the ages of 40 and 60, most have no significant health consequences. Variations in health are related to genetic, educational, and economic factors, as well as gender. The most important reason for individual variations in health, however, is personal lifestyle.

In the audio program, "Improving the Odds," we meet two middle-aged individuals concerned about the health of their changing bodies. As the program unfolds, Susan, 47, and Larry, 59, complete the **life-expectancy questionnaire** designed by psychologist Diana Woodruff-Pak. Although the test obviously cannot predict how long each person will live, it is a useful tool that identifies factors likely to extend or shorten their lives. These factors focus on each person's **genetic history, personal health habits, socioeconomic status,** and **social and personality characteristics.** As Susan and Larry work through the questionnaire, it becomes clear that certain factors related to longevity, such as how long one's ancestors lived, are beyond the individual's control. But factors under one's control can add or subtract 20 years from a life. We wonder what the questionnaire predicts for Larry and Susan.

LESSON GOALS

By the end of this lesson, you should be prepared to:

1. Describe the typical pattern of physical development that occurs during middle adulthood.

2. Explain the relationship of an individual's life expectancy to lifestyle, genetic history, health practices, and personality.

Audio Assignment

Listen to the audio tape that accompanies Lesson 20: "Improving the Odds." Write answers to the following questions. You may replay portions of the program if you need to refresh your memory. Answer guidelines may be found in the section at the end of this chapter.

1. What is the difference in life expectancy for men and women and why do researchers believe that this difference is a biological rather than a social phenomenon?

2. What is the relationship between an individual's predicted longevity and his or her **genetic history?**

3. In what ways do **personal health habits** predict longevity?

4. What is the relationship between an individual's predicted longevity and his or her **socioeconomic status?**

5. What is the relationship between an individual's predicted longevity and his or her **social** and **personality characteristics?**

6. What does it mean when there is a **correlation** between two variables?

7. Several limitations of the life expectancy test were mentioned in the program. What are they?

Textbook Assignment

Read Chapter 20: "Middle Adulthood: Physical Development," pages 463–483 in *The Developing Person Through the Life Span, 2/e.*

Write your answers to the following questions. Refer back to the textbook if necessary. Answer guidelines may be found in the Lesson Guidelines section at the end of the chapter.

1. What are some of the signs of aging that occur in middle-aged adults?

2. What are some of the ways in which stress influences health?

3. What are some of the typical changes that occur in the sexual–reproductive system during middle adulthood?

Testing Yourself

After you have completed the audio and text review questions, see how well you do on the following quiz. Correct answers, with text and audio references, may be found at the end of this chapter.

1. The life expectancy advantage that females have over males is:
    a. found in no species other than the human.
    b. probably a social rather than a biological effect.
    c. probably a biological rather than a social effect.
    d. only a recent historical development.

2. Individuals who consume diets rich in _____ tend to live the longest.
    a. vegetables, fruits, and simple foods
    b. meat, fish, and other high-protein foods
    c. saturated fats
    d. simple carbohydrates

3. Which of the following is *not* a characteristic of Type A behavior?
    a. Achievement striving
    b. Impatience and aggressiveness
    c. Having more friends than people who do not show the Type A pattern
    d. Increased adrenalin and blood pressure levels in response to stress

4. Research discussed in the text indicates that as people grow older:
    a. they seem to experience less stress, or to be better at coping adaptively with stress.
    b. they usually experience greater stress than younger adults.
    c. the stressors they encounter are about the same, but their ability to cope adaptively declines.
    d. they react to stress just as they did when they were younger.

5. On the average in the United States, the socioeconomic status of black persons is _____ than that of white persons. This may explain why the life expectancy of blacks is _____ than that of whites.
    a. higher . . . longer
    b. higher . . . shorter
    c. lower . . . longer
    d. lower . . . shorter

6. People who live in _____ tend to live longer than people who live in _____.
    a. the south . . . the north
    b. the north . . . the south
    c. urban areas . . . rural areas
    d. rural areas . . . urban areas

7. Studies of menopausal women reported in the textbook suggest that:
    a. virtually all find menopause to be an extremely unpleasant time in their lives.
    b. many find menopause to be a liberating and more benign experience than its reputation suggests.
    c. today, menopause is occurring at an earlier average age than ever before.
    d. today, menopause is occurring later in life than ever before.

8. The most frequent cause(s) of death in the United States is (are):
    a. cancer.
    b. accidents.
    c. cardiovascular disease.
    d. acute illness.

9. At age 80 there are about twice as many women as men alive; this is probably due to the fact that:
    a. although males are born at a higher rate than are females, they are more susceptible to the hazards of life at every age.
    b. estrogen protects females against the cardiovascular diseases that kill males.
    c. testosterone makes males more susceptible to cardiovascular disease.
    d. all of the above are true.

10. The major influence on longevity is an individual's:
    a. personal health habits.
    b. socioeconomic status.
    c. genetic history.
    d. personality.

11. The climacteric refers to:
    a. the various biological and physiological changes that accompany menopause.
    b. the increased life expectancy that has occurred in the past 50 years.
    c. the psychological stress of aging.
    d. the decreased frequency of sexual intercourse that typically occurs during middle adulthood.

12. According to the text, the overall impact of aging on the individual depends in large measure on the individual's:

    a. socioeconomic status.
    b. sex.
    c. attitude.
    d. health.

13. Over the past 35 years, the death rate from heart disease and cancer among middle-aged people has:

    a. increased.
    b. not changed significantly.
    c. decreased for men and increased for women.
    d. decreased.

14. According to the text, which of the following is considered the number one health problem in the United States?

    a. Cigarette smoking
    b. Poor nutrition
    c. Alcohol abuse
    d. Obesity

15. The effects of stress on the body include:

    a. increased heart rate.
    b. increased blood pressure.
    c. suppression of the immune system.
    d. all of the above.

LESSON 20 EXERCISE: HOW LONG WILL YOU LIVE?

An individual's life span is determined by many factors, including genetic history, personal health habits, socioeconomic status, and personality. To see how these factors interact, complete the following life-expectancy questionnaire for yourself, or for someone you know. Begin by looking at the life-expectancy table (page 259) to find the number of years an average individual of your age, gender, and race is expected to live. Then, as you check through the list, add or subtract the appropriate number of years for each item.*

Beginning Life Expectancy (from Table) _____ *80*

1. **Longevity of grandparents**
   Add 1 year for each grandparent living beyond age 80. Add one-half year for each grandparent surviving beyond the age of 70. _____ *+1*

2. **Longevity of parents**
   If your mother lived beyond the age of 80, add 4 years. Add 2 years if your father lived beyond 80. _____ *0*

3. **Cardiovascular disease among close relatives**
   If any parent, grandparent, or sibling died from cardiovascular disease before age 50, subtract 4 years for each incidence. If any died from the above before the age of 60, subtract 2 years. _____ *0*

* Abridged with permission from Diana Woodruff-Pak (1977). *Can you live to be one hundred?* New York: Chatham Square Press.

4. Other heritable disease among close relatives
If any parent, grandparent, or sibling died before the age of 60 from diabetes or peptic ulcer, subtract 3 years. If any died before 60 from stomach cancer, subtract 2 years. Women whose close female relatives have died before 60 from breast cancer should also subtract 2 years. Finally, if any close relatives have died before the age of 60 from any cause except accidents or homicide, subtract 1 year for each incidence.   _-1_

5. Childbearing
Women who cannot, or do not plan to have children, and those over 40 who have never had children, should subtract one-half year. Women who have had over seven children, or plan to, should subtract 1 year.   _0_

6. Mother's age at your birth
Was your mother over the age of 35 or under the age of 18 when you were born? If so, subtract 1 year.   _-1_

7. Birth order
Are you the first born in your family? If so, add 1 year.   _0_

8. Intelligence
If you feel that you are superior in intelligence add 2 years.   _+2_

9. Weight
If you are more than 30 percent overweight, subtract 5 years. If you are more than 10 percent overweight, subtract 2 years.   _0_

10. Dietary habits
If you eat a lot of vegetables and fruits, and usually stop eating before feeling full, add 1 year. If you drink five or more cups of coffee per day, subtract one-half year.   _0_

11. Smoking
If you smoke two or more packs of cigarettes a day, subtract 12 years. If you smoke between one and two packs a day, subtract 7 years. If you smoke less than a pack a day, subtract 2 years.   _-2_

12. Drinking
If you are a moderate drinker, add 3 years. If you are a light drinker, add 1.5 years. If you are a heavy drinker, subtract 8 years.   _+3_

13. Exercise
If you exercise briskly at least three times a week, add 3 years.   _0_

14. Sleep
If you sleep more than 10 hours or less than 5 hours a night, subtract 2 years.   _0_

15. Sexual activity
If you enjoy sexual activity at least once a week, add 2 years.   _+2_

16. Regular physical examinations
If you have an annual physical examination by your physician, add 2 years.   _+2_

17. **Health status**
    If you have a chronic illness at present, subtract 5 years.

    0 _____

18. **Years of education**
    If you graduated from college, add 4 years. If you attended college but did not graduate, add 2 years. If you graduated from high school but did not attend college, add 1 year. If you have less than an eighth-grade education, subtract 2 years.

    +2 _____

19. **Occupational level** (former, if retired; spouse's, if you are not working)
    Professional, add 1.5 years; technicians, administrators, managers, and agricultural workers, add 1 year; semi-skilled workers should subtract one-half year; laborers should subtract 4 years.

    0 _____

20. **Family income**
    If your family income is above average for your education and occupation, add 1 year. If it is below average for your education and occupation, subtract 1 year.

    0 _____

21. **Activity on the job**
    If your job involves a lot of physical activity, add 2 years. If your job requires that you sit all day, subtract 2 years.

    +2 _____

22. **Age and work**
    If you are over the age of 60 and still on the job, add 2 years.
    If you are over the age of 65 and have not retired, add 4 years.

    0 _____

23. **Rural vs. urban dwelling**
    If you live in an urban area and have lived in or near the city for most of your life, subtract 1 year. If you have spent most of your life in a rural area, add 1 year.

    +1 _____

24. **Married vs. divorced**
    If you are married and living with your spouse, add 1 year.

    Men: If you are separated or divorced and living alone, subtract 9 years (not alone: subtract 4 years). If you are widowed and living alone subtract 7 years (not alone: subtract 3 years).

    0 _____

    Women: If you are separated or divorced and living alone, subtract 4 years. If you are widowed and living alone, subtract 3 years. If you are separated, divorced, or widowed and not living alone, subtract 2 years.

    0 _____

25. **Single living status**
    Unmarried women (living alone or with others) and unmarried men who live with family or friends should subtract 1 year for each unmarried decade past age 25. Unmarried men who live alone should subtract 2 years for each decade after 25.

    0 _____

26. **Life changes**
    If you are always changing things in your life—jobs, residences, friends—subtract 2 years.

    0 _____

27. **Friendship**
    If you have at least two close friends in whom you can confide almost all the details of your life, add 1 year.

    +1 _____

28. **Aggressive personality**
    If you have an aggressive and sometimes hostile personality,
    subtract 2 years.

    −0

    _____

29. **Flexible personality**
    If you are a calm, easygoing, adaptable person, add 2 years. If you
    are rigid, dogmatic, and set in your ways, subtract 2 years.

    +2

    _____

30. **Risk-taking personality**
    If you take a lot of risks, including driving without seat belts, exceeding the
    speed limit, and taking any dare that is made, subtract 2 years. If you use
    seat belts regularly, drive infrequently, and generally avoid risks
    and dangerous parts of town, add 1 year.

    +1

    _____

31. **Depressive personality**
    Have you been depressed, tense, worried, or guilty for more than a period
    of a year or two? If so, subtract 1 to 3 years depending upon how
    seriously you are affected by these feelings.

    0

    _____

32. **Happy personality**
    Are you basically happy and content, and have you had a lot of
    fun in life? If so, add 2 years.

    +2

    _____

After you have completed the longevity questionnaire, fill in the information
requested on the **Exercise Response Sheet**. Send the completed response sheet
to your instructor.

Life Expectancy Table

Your Age in 1983	White		Black	
	Male	Female	Male	Female
10	72.7	79.6	67.2	75.2
12	72.7	79.6	67.2	75.2
14	72.8	79.6	67.3	75.2
16	72.9	79.7	67.4	75.3
18	73.0	79.7	67.5	75.3
20	73.1	79.8	67.6	75.5
22	73.3	79.8	67.8	75.5
24	73.4	79.9	68.0	75.6
26	73.6	80.0	68.2	75.7
28	73.7	80.0	68.4	75.8
30	73.9	80.1	68.7	75.9
32	74.0	80.1	69.0	76.0
34	74.1	80.2	69.3	76.2
36	74.3	80.3	69.6	76.3
38	74.4	80.3	69.9	76.5
40	74.6	80.4	70.2	76.7
42	74.7	80.5	70.6	76.9
44	74.9	80.7	71.0	77.1
46	75.1	80.8	71.4	77.4
48	75.4	81.0	71.9	77.7
50	75.7	81.2	72.5	78.1
52	76.0	81.4	73.1	78.5
54	76.4	81.7	73.7	78.9
56	76.8	81.9	74.4	79.4
58	77.3	82.3	75.2	79.9
60	77.9	82.6	76.0	80.5
62	78.5	83.0	76.9	81.2
64	79.2	83.5	77.9	81.9
66	79.9	84.0	78.9	82.6
68	80.7	84.5	79.9	83.3
70	81.5	85.1	80.9	84.1
72	82.5	85.7	82.1	85.0
74	83.4	86.4	83.3	86.0
76	84.5	87.2	84.6	87.0
78	85.7	88.0	81.8	87.9
80	86.9	88.8	87.1	89.0
82	88.2	89.8	88.5	90.2
84	89.5	90.9	90.1	91.6

Source: Vital Statistics of the United States.

LESSON GUIDELINES

Audio Question Guidelines

1. In all species of animals in the wild, the female of the species lives longer. The fact that this is true of all species, including humans, suggests a biological rather than a social effect.

   For humans, women maintain this biological edge from the first instant of life to the very end. About 120 males are conceived for every 100 females, but the hazards of male development are so great that at birth the ratio is down to 106 males for every 100 females. At age 80 there are only 50 males for every 100 females alive.

   Women are probably protected from cardiovascular disease (the most common cause of death in the United States) by the hormone estrogen. Because of their higher testosterone levels, men may be at increased risk for cardiovascular disease.

2. Genetic history has the major influence on an individual's longevity. In order of importance, the longevity of one's mother, father, and grandparents is correlated with one's own life expectancy.

3. In promoting longevity, the following personal health habits are important: maintaining ideal and stable body weight; eating a balanced diet rich in fruits, vegetables, and simple foods, yet low in fat and sugar; not smoking; drinking alcohol in moderation; engaging in regular exercise; and having regular physical examinations.

4. The higher a person's socioeconomic status, the longer he or she is likely to live. Longevity tends to be greater in people who have had more education, those who work in professional or managerial (rather than unskilled) professions, and in those with above average income for their age and occupation.

   The lower average socioeconomic status of blacks in the United States may explain why their life expectancy is approximately 6 years less than that of white persons. Persons with low socioeconomic status are more likely to live in conditions less conducive to the maintenance of good health.

5. Greater longevity is found more commonly among persons who are socially integrated than among those who are not.

   Especially among men, divorce, widowhood, being single, and separation predict shorter life expectancy.

   Experiencing lots of changes in one's life is also associated with reduced life expectancy.

   Being a happy person generally and having at least two close friends predicts a longer life expectancy.

   Being an aggressive personality is associated with reduced life expectancy.

6. When two factors or variables are correlated, it means that changes in one are predictive of changes in the other. A correlation between two variables, however, does not imply that changes in one *cause* changes in the other. A

third factor might influence the two variables, which, although linked in a correlational fashion, do not influence one another.

7. Because the life-expectancy test is an example of correlational, rather than experimental, research, no conclusions about what "causes" a person to have lengthened or reduced life expectancy can be drawn.

The life-expectancy test relies on the subjective assessment and memory of the respondent. The test's validity and accuracy are therefore, subject to question.

There has been no longitudinal "follow-up" of individuals who have taken the test to see if the predictions were accurate.

Textbook Question Guidelines

1. The physical signs of middle age include graying and thinning of the hair; drying and wrinkling of the skin; change in body shape as pockets of fat settle on the upper arms, buttocks, and eyelids; loss of height as back muscles, connecting tissues, and bones lose strength; and an increase in the number of overweight individuals.

Each of the senses becomes less acute during middle age, with decline occurring at about the same rate in hearing, vision, and the other senses.

2. Whether an experience is stressful depends on the particular individual; how that person responds physiologically and behaviorally; the individual's temperament, past experiences, physical vulnerabilities, resources, and strategies for coping; the overall context; and how these factors influence the way the individual interprets a potential stressor.

Virtually every disease becomes more likely in people who experience significant stress. Stress also elevates heart rate and blood pressure, and suppresses function of the immune system.

Type A persons have especially maladaptive physiological responses to stress.

3. After about age 35, many women experience a reduction in the time between menstrual periods. Sometime between their late 40s and early 50s, most women reach **menopause,** during which ovulation and menstruation stop and the production of estrogen drops significantly. The **climacteric** refers to the various physiological changes that accompany menopause.

The prevalent conception of menopause as a time of great difficulty and depression is largely a myth.

Physiologically, for men, there is no sudden shift in hormone levels or reproductive ability, although age-related declines in the number and motility of sperm do occur. Testosterone levels may drop, however, if a man becomes sexually inactive, or experiences unusual anxiety.

In terms of the frequency of intercourse and orgasm, sexual activity usually declines during middle age. Sexual stimulation, especially in men, takes longer and needs to be more direct than when they were younger.

For middle-aged and older adults, present enjoyment and interest in sex is more strongly correlated with past enjoyment and interest than with other physiological variables.

Answers to Testing Yourself

1. **c.** In all species of animals studied, including the human, the female of the species lives longer than the male. (audio program)

2. **a.** People who live the longest are not overly concerned with food, but tend to eat diets that emphasize vegetables, fruits, and simple foods. (audio program)

3. **c.** Type A persons tend to have fewer friends and lower levels of happiness than those who do not exhibit the Type A pattern. (textbook, p. 475)

4. **a.** Several studies have found that as people grow older, they seem to experience less stress or become better at coping adaptively with stress. (textbook, pp. 471–477)

5. **d.** The life expectancy of black persons in the United States is about 6 years less than that of white persons. This is probably due to the lower socioeconomic status of blacks, which often correlates with less healthful living conditions. (audio program)

6. **d.** The advantage is small, but people who live in rural areas do tend to live longer than those who live in urban areas. (audio program)

7. **b.** The prevalent conception of menopause as a time of difficulty and depression is largely a myth. (textbook, pp. 478–481)

8. **c.** Cardiovascular disease is the leading cause of death in the United States. (textbook, pp. 466–470; audio program)

9. **d.** All of the above are true. (audio program)

10. **c.** Personality, health habits, and socioeconomic status are important, but personal genetic history is the major influence on how long a person will live. (audio program)

11. **a.** The climacteric includes vasomotor instability, drier skin, less vaginal lubrication, and other physiological changes that accompany menopause. (textbook, p. 478)

12. **c.** The impact of aging on a person depends in large measure on his or her attitude. (textbook, pp. 463–465)

13. **d.** Reflecting improved health in this cohort, the overall death rate among the middle-aged—especially from heart disease and cancer—has declined dramatically over the past 35 years. (textbook, pp. 467–469)

14. **c.** Alcohol abuse is considered the number one health problem in the United States. (textbook, p. 468)

15. **d.** Increased heart rate, elevated blood pressure, and a suppression of immune function are all effects of stress of the body. (textbook, pp. 471–474)

References

Woodruff, D. (1977). *Can you live to be one hundred?* New York: Chatham Square Press.

Middle Adulthood: Cognitive Development

AUDIO PROGRAM: **What Makes an Expert?**

ORIENTATION

For most of this century, psychologists were convinced that intelligence peaks during adolescence and then gradually declines throughout adulthood. Within the past 25 years, however, research has led to the opposite conclusion, that in some ways intelligence actually improves during adulthood. Audio program 21 explores how intelligence changes through the adult years and describes the methodology by which developmental psychologists study these changes.

Chapter 21 of the textbook notes that researchers today believe that there are several kinds of intelligence, each of which may increase, decrease, or remain stable with age. Some experts maintain that **fluid intelligence,** based on the underlying abilities of short-term memory, abstract thought, and speed of thinking, declines with age while **crystallized intelligence,** based on accumulated general knowledge, increases. Each dimension of intelligence follows its own developmental pattern, which is determined in part by the individual's education and life experiences, and by **cohort,** or generational, differences. In fact, during middle adulthood **individual variation** is probably more important in influencing cognitive development than is chronological age.

The audio program, "What Makes an Expert?" states that as people grow older they get better and better at the things that are important to them, while abilities that are not practiced decline. Focusing on the particular **expertise** of a musical savant and a professor of surgery, the program explores the many ways in which experts are better than novices at what they do. They are more intuitive and flexible, use better problem-solving strategies, and often process information and perform automatically. During the program, commentary is provided by psychologist Neil Charness and Professor of Surgery George Zuidema.

As the program opens we hear a piano sonata played by John LaFond. Although he has been blind since birth, suffers from severe epilepsy, is mentally retarded, and nearly paralyzed on the right side of his body, LaFond has specialized very successfully in one domain: music.

LESSON GOALS

By the end of this lesson, you should be prepared to:

1. Discuss the controversy surrounding changes in adult intelligence and the role of research methods in fueling that controversy.

2. Explain contemporary views of intelligence and contrast them with those put forward by earlier theories.

3. Discuss the complexity and plasticity of the development of intelligence during the adult years.

4. Describe the distinguishing features of expertise.

Audio Assignment

Listen to the audio tape that accompanies Lesson 21: "What Makes an Expert?"
Write answers to the following questions. You may replay portions of the program if you need to refresh your memory. Answer guidelines may be found in the Lesson Guidelines section at the end of this chapter.

1. Explain what the abilities of a middle-aged musical savant, a chess grand master, and a skilled physician indicate about the nature of intelligence.

2. Describe the ways in which thinking changes as a person develops expertise in a particular area.

3. Discuss whether experts in different fields of specialization have different peak years of achievement and productivity during the life span.

Textbook Assignment

Read Chapter 21, "Middle Adulthood: Cognitive Development," pages 485–497 in *The Developing Person Through the Life Span, 2/e.*
Write your answers to the following questions. Refer back to the textbook, if necessary. Answer guidelines may be found in the Lesson Guidelines section at the end of this chapter.

1. What evidence from research led to the widespread belief that intelligence declines during adulthood?

2. What evidence suggests that intelligence increases during adulthood?

3. Differentiate between fluid and crystallized intelligence and explain how each is affected by age.

4. Describe the contemporary view of intelligence, which places emphasis on the multidimensional and multidirectional nature of cognitive abilities.

5. Discuss the impact of individual differences, plasticity, and encapsulation on intelligence.

Testing Yourself

After you have completed the audio and text review questions, see how well you do on the following quiz. Correct answers, with text and audio references, may be found at the end of this chapter.

1. Research on expertise indicates that during adulthood, intelligence:
   a. increases in most of the primary mental abilities.
   b. increases in specific areas of interest to the person.
   c. increases only in those areas associated with the individual's career.
   d. shows a uniform decline in all areas.

2. John LaFond, a musical savant, can easily reproduce a piano melody that he has heard for the first time. His ability to do so demonstrates expertise based on:
   a. superior short-term memory for individual notes.
   b. superior working memory for individual notes.
   c. superior ability to recognize and remember familiar musical patterns.
   d. compensation for retardation in other areas.

3. Which of the following is *not* characteristic of expertise, as described in the audio program?
   a. An intuitive approach to performance
   b. Automatic cognitive processing
   c. A heightened ability to recognize familiar patterns *and* unusual cases
   d. Superior intelligence and intellectual functioning

4. Compared to the peak years for achievement in mathematics, the peak years for achievement in history:
   a. tend to come at an earlier age.
   b. tend to come at a later age.
   c. tend to come at about the same age.
   d. cannot be predicted with any degree of accuracy.

5. The belief that intelligence always declines during adulthood was based on:
   a. cross-sectional research.
   b. longitudinal research.
   c. sequential research.
   d. a multidimensional model of intelligence.

6. The results of Schaie's study of adults between the ages of 25 and 67 indicated that:
   a. younger adults performed better than older adults for all mental abilities.
   b. intelligence in all primary mental abilities actually increased with each decade.
   c. for some abilities, older adults performed better than younger adults.
   d. crystallized, but not fluid, intelligence decreased with each decade.

7. A psychologist has found significant differences in linguistic intelligence in adults born in the 1920s and those born in the 1950s. She suspects that this difference is a reflection of different educational emphases of the two historical periods. This is an example of:
   a. the advantage of longitudinal research.
   b. sequential research.
   c. a cohort difference.
   d. all of the above.

8. Horn's research indicates that during adulthood, declines occur in:
   a. crystallized intelligence.
   b. fluid intelligence.
   c. both crystallized and fluid intelligence.
   d. neither crystallized nor fluid intelligence.

9. Sharetta knows more about her field of specialization now at age 45 than she did at age 35. This increase is most likely due to:
   a. an increase in crystallized intelligence.
   b. an increase in fluid intelligence.
   c. increases in both fluid and crystallized intelligence.
   d. a cohort difference.

10. Fluid intelligence is based on all of the following *except:*
    a. short-term memory.
    b. abstract thinking.
    c. speed of thinking.
    d. general knowledge.

11. In recent years, researchers are more likely than before to consider intelligence as:
   a. a single entity.
   b. primarily determined by heredity.
   c. entirely the product of learning.
   d. made up of several abilities.

12. Among adults, the *best* predictor of individual differences in intelligence is:
   a. age.
   b. ethnicity.
   c. interindividual variation and cohort differences.
   d. socioeconomic status.

13. Which of the following is the best example of encapsulation?
   a. As Yusef has grown older his knowledge of history, geography, and general information has increased tremendously.
   b. Although Kim is a leading authority on geology, she knows little about sociology, economics, or even other scientific disciplines.
   c. Duane, a brain-damaged idiot savant, is gifted in manipulating numbers but is otherwise severely retarded.
   d. Susan, a successful realtor, has written a popular book on the history of landscaping.

14. One of the drawbacks of longitudinal studies of intelligence is that:
   a. they are especially prone to the distortion of cohort effects.
   b. people who are retested may show improved performance as a result of practice effects.
   c. the biases of the experimenter are more likely to distort the results than occurs with other research methods.
   d. all of the above are probably true.

15. A contemporary developmental psychologist is most likely to *disagree* with the statement that:
   a. many people show increases in intelligence during middle adulthood.
   b. for many behaviors, the responses of older adults are slower than those of younger adults.
   c. intelligence peaks during adolescence and declines thereafter.
   d. intelligence is multidimensional and multidirectional.

LESSON 21 EXERCISE: CREATIVITY

One theme of this lesson is that contemporary psychologists take a broader view of intelligence than was the case in previous years. Experts recognize that earlier studies of intelligence failed to consider that generational differences, or **cohort effects,** may influence scores on standardized intelligence tests. Intelligence is now considered multidimensional and multidirectional in nature rather than being a single, fixed entity. One dimension of intelligence is the specialized knowledge that comes with the development of expertise. Another is creativity. The term "creativity" is used to describe the behavior of individuals who are able to find novel, and practical, solutions to problems.

How is creativity related to more traditional dimensions of intelligence? Research has shown that although a certain degree of intelligence is obviously necessary for creativity to be manifest, other factors, such as individual life experiences, are also important.

In attempting to study how creativity changes during adulthood, developmentalists have used several approaches. On the following page is a copy of the *Remote Associates Test* devised by Sarnoff and Mednick. This test is based on the idea that creativity reflects an ability to see relationships among ideas that are remote from one another. Several studies have reported that creative abilities tend to hold up well through middle adulthood, and may even extend well into late adulthood. This is especially true for individuals who regularly engage in creative thinking, such as those whose professions require and call upon their creativity.

Arrange to administer the *Remote Associates Test* to two individuals, preferably a young adult or adolescent, and an older adult. If you wish to take the test yourself, do so first, and then test your other subject. Two copies of the test are printed, one for each of your subjects. Instructions for the test are given on the test sheet. You will need to time the number of minutes it takes for you and/or your subject(s) to complete the test. Correct answers to the test are given at the end of this chapter following the Lesson Guidelines section. When you have finished the testing, complete the **Exercise Response Sheet** and return it to your instructor.

REMOTE ASSOCIATES TEST

Instructions: In this test you are presented with three words and asked to find a fourth word that is related to all three. Write this word in the space to the right.
For example, what word do you think is related to these three?

paint   doll   cat   _____   930

The answer in this case is "house": house paint, doll house and house cat.

#			
1. call	pay	line	*phoNE*
2. end	burning	blue	*caNDLE*
3. man	hot	sure	*FiRE*
4. stick	hair	ball	*piN*
5. blue	cake	cottage	*CHEESE*
6. man	wheel	high	
7. motion	poke	down	*SloE*
8. stool	powder	ball	*Foot*
9. line	birthday	surprise	*PaRty*
10. wood	liquor	luck	*haRD*
11. house	village	golf	
12. plan	show	walker	
13. key	wall	precious	
14. bell	iron	tender	
15. water	pen	soda	
16. base	snow	dance	*BaLL*
17. steady	kart	slow	
18. up	book	charge	
19. tin	writer	my	
20. leg	arm	person	
21. weight	pipe	pencil	
22. spin	tip	shape	
23. sharp	thumb	tie	
24. out	band	night	
25. cool	house	fat	
26. back	short	light	
27. man	order	air	*MaiL*
28. bath	up	gum	*BUBBLE*
29. ball	out	jack	
30. up	deep	rear	

Source: Gardner, R. (1980). *Exercises for general psychology.* Minneapolis: Burgess, 115–116.

REMOTE ASSOCIATES TEST

Instructions: In this test you are presented with three words and asked to find a fourth word that is related to all three. Write this word in the space to the right.
For example, what word do you think is related to these three?

paint   doll   cat   _____

The answer in this case is "house": house paint, doll house and house cat.

1. call   pay   line   _____
2. end   burning   blue   _____
3. man   hot   sure   _____
4. stick   hair   ball   _____
5. blue   cake   cottage   _____
6. man   wheel   high   _____
7. motion   poke   down   _____
8. stool   powder   ball   _____
9. line   birthday   surprise   _____
10. wood   liquor   luck   _____
11. house   village   golf   _____
12. plan   show   walker   _____
13. key   wall   precious   _____
14. bell   iron   tender   _____
15. water   pen   soda   _____
16. base   snow   dance   _____
17. steady   kart   slow   _____
18. up   book   charge   _____
19. tin   writer   my   _____
20. leg   arm   person   _____
21. weight   pipe   pencil   _____
22. spin   tip   shape   _____
23. sharp   thumb   tie   _____
24. out   band   night   _____
25. cool   house   fat   _____
26. back   short   light   _____
27. man   order   air   _____
28. bath   up   gum   _____
29. ball   out   jack   _____
30. up   deep   rear   _____

Source: Gardner, R. (1980). Exercises for general psychology. Minneapolis: Burgess, 115–116.

LESSON GUIDELINES

Audio Question Guidelines

1. LaFond's ability in music (compared to his severe general retardation) indicates that intelligence can be specialized very narrowly. LaFond's memory span for individual notes is not unusually high. Rather, as a result of spending thousands of hours at the piano, LaFond has developed an uncanny ability to recognize and remember familiar *patterns* of notes.

   This superior pattern memory is similar to the "Grand Master intuition" seen in expert chess players. After many years of experience, chess masters have built up a large memory repertoire of chess patterns that helps them to play more intuitively, recognize instantly the structure of a situation, and determine its likely outcome.

   Experienced physicians diagnose symptoms more quickly and spot rare cases as a result of recognizing familiar patterns.

2. Relying more than novices on their accumulated experience, experts are more intuitive and less stereotyped in their problem-solving behaviors.

   Many elements of expert performance become automatic and less tied to focused attention.

   As expertise is acquired, certain skills and cognitive processes become more specialized.

   Experts generally have more, and better, strategies for accomplishing particular tasks.

   Experts tend to be more flexible in their work.

3. Most people tend to do their greatest work in the decade of their 30s. This varies from field to field, however.

   In fields such as mathematics, the peak years of achievement tend to be a little earlier. The individual needs fewer facts before he or she can go to work and be productive.

   In fields such as history, which require the accumulation of a greater knowledge base, the peak years of achievement tend to be somewhat later.

Textbook Question Guidelines

1. Researchers are divided on the issue of whether intelligence declines as aging occurs. For most of this century, researchers reported that intellectual ability peaks between ages 18 and 21 and then declines steadily with age.

   This evidence is based on many large **cross-sectional** studies, in which groups of people who are different in age are compared.

   Cross-sectional studies ignore **cohort differences** (generational effects). This is a serious shortcoming of this type of research.

2. **Longitudinal studies,** in which the same people are evaluated over a long period of time, have tended to show that most intellectual abilities increase throughout adulthood.

One drawback of longitudinal research is that being retested several times on similar items might improve a person's performance.

The **sequential research** method is designed to correct for the limitations of cross-sectional and longitudinal research methodologies. Each time the original group of subjects is retested (longitudinal design), a new group of adults is tested, thus controlling for the possible effects of retesting as well as revealing the impact of cohort differences.

Sequential research has found that adults in their 30s and 40s score higher on tests of many primary abilities than adults in their 20s. Not until subjects reach their 70s are the scores lower than the average scores of 25-year-olds.

3. **Fluid intelligence** is made up of basic mental abilities such as short-term memory, abstract thinking, and speed of thinking.

Crystallized intelligence refers to the accumulation of facts, information, and learning strategies that comes with education and experience.

Although it was once believed that fluid intelligence was inherited and crystallized intelligence was learned, this nature–nurture distinction is probably not valid.

There is some research evidence that fluid intelligence declines during adulthood while crystallized intelligence increases.

4. In the past, psychologists viewed intelligence as a single entity. Recently, several researchers have proposed a **multidimensional** view of intelligence, arguing that there are many distinct cognitive abilities, each of which follows its own developmental path depending on the person's age, life experiences, interests, and education.

Gardner, for example, proposes seven autonomous intelligences: linguistic, musical, logical–mathematical, spatial, bodily–kinesthetic, self-understanding, and social understanding.

Contemporary researchers also believe that since intellectual abilities are **multidirectional,** it is misleading to ask whether intelligence in general increases or decreases through the life span.

5. Intellectual development is greatly influenced by individual differences and cohort differences. Some individuals seem to decline in all mental abilities during middle adulthood, while others are just as capable at 70 as they were at a younger age.

Contemporary researchers also emphasize the **plasticity** of intelligence. Instead of their being innate and immutable, the development of intellectual abilities is shaped by an individual's experiences.

Some researchers have suggested that as we age, our intelligence increases in the specific areas that are important to us.

Encapsulation refers to the concentration of intellectual strengths in one small area. Through encapsulation, the individual may become an expert in one area, yet be average or below average in knowledge about other areas.

Answers to Testing Yourself

1. **b.** The widespread belief that intelligence inevitably declines during adulthood is based on a misconception of intelligence as a single, fixed entity rather than a multidimensional and multidirectional entity. (audio program)

2. **c.** Experts do not have "better" memory per se; the key to their expertise lies in how their knowledge and memories are organized.(audio program)

3. **d.** There is no evidence that experts are more "intelligent" than nonexperts; moreover, the concept of intelligence as a single general ability is probably not valid. (audio program)

4. **b.** Success in fields such as history is based partly on the accumulation of knowledge over a long period of time. (audio program)

5. **a.** Because it is impossible to select adults who are similar to each other in every aspect except age, the results of cross-sectional research are often distorted by cohort differences. (textbook, pp. 487–488)

6. **c.** Schaie's studies indicated that most people improved in primary mental abilities during most of adulthood. (textbook, pp. 487–488)

7. **c.** Each generation, or cohort, has its own characteristics. (textbook, p. 487)

8. **b.** The decline in fluid intelligence is temporarily masked by an increase in crystallized intelligence during adulthood. (textbook, p. 490)

9. **a.** Crystallized intelligence is the accumulation of facts, information, and learning strategies that comes with education and experience. (textbook, pp. 489–490)

10. **d.** Knowledge of general facts is the basis of crystallized intelligence. (textbook, p. 489)

11. **d.** The closer researchers look at adult intelligence, the clearer it becomes that intelligence is multidimensional. (textbook, pp. 492–493)

12. **c.** One reason that there are many patterns in the development of intelligence is that each individual is genetically unique and has a unique repertoire of experiences. (textbook, p. 493)

13. **b.** During adulthood our intelligence increases in very specific areas, through encapsulation, by which our intellectual strengths become concentrated in a limited area. (textbook, p. 494)

14. **b.** Such practice effects are a serious drawback of the longitudinal method. (textbook, p. 491)

15. **c.** The belief that intelligence inevitably declines during adulthood is based on a misconception of intelligence as a single, rather than multidimensional entity, and as fixed, rather than flexible. (audio program; textbook, pp. 492–494)

Answers to the Remote Associates Test

1. phone 2. book 3. fire 4. pin 5. cheese 6. chair 7. slow 8. foot
9. party 10. hard 11. green 12. floor 13. stone 14. bar 15. fountain
16. ball 17. go 18. cover 19. type 20. chair 21. lead 22. top 23. tack
24. watch 25. cat 26. stop 27. mail 28. bubble 29. black 30. end

References

Charness, N. (1986). Expertise in chess, music, and physics: A cognitive perspective. In L.K. Obler and D.A. Fein (Eds.), *The neuropsychology of talent and special abilities*. New York: Guilford Press.

> Professor Charness, who is heard on the audio program, discusses further distinctions that can be made between those who are experts and those who are less skilled.

Schaie, K.W., and Herzog, C. (1983). Fourteen-year cohort–sequential studies of adult intelligence. *Developmental Psychology, 19*, 531–543.

> Professor Schaie, who was one of the earliest researchers to recognize the distorting effects of cohort differences, discusses issues raised by the assessment of cognitive development during adulthood.

Middle Adulthood: Psychosocial Development

AUDIO PROGRAM: The Life Course of Work

ORIENTATION

Lesson 22 is concerned with middle age, a period when the reevaluation of career goals, shifts in one's family responsibilities, and a growing awareness of one's mortality often lead to turmoil and change. Chapter 22 of *The Developing Person Through the Life Span, 2/e,* explores several issues concerning development during middle age. The first is the so-called **midlife crisis.** Is it fact or fantasy, and do women feel it as much as men? The second issue regards the changing relationships of middle-aged adults with their adult children and aging parents. Being "sandwiched" between the younger and older generations is often a source of stress, as middle-aged adults experience additional financial, emotional, and caregiving demands.

The chapter also explores career development during middle age—the topic of audio program 22, "The Life Course of Work." Most of us define ourselves by the work we perform. But what happens when the basic structure of our job changes? Or we change? At some point during middle age, most adults reach a plateau in their career development that prompts a reevaluation of career objectives. For some workers, midlife may bring a period of **burn-out** or **alienation.** Many may have to modify their dreams and make the best of their situation.

In the program, the stories of Mary and Dan illustrate how the life course of work has changed. Mary, 48, returned to college when the youngest of her three children started high school. She has continued with graduate training in the hopes of beginning a new career. Sociologist Alice Rossi, who has done extensive research on the work and family lives of women, offers a historical perspective on women like Mary, who return to school and then to work after careers as mothers and homemakers.

Dan, 56, has practiced dentistry for 30 years. Although he would never think of quitting his profession, changes in the field have caused him to stop recommending it to others as a career. Professor of Business Stephen Lazarus, who has worked extensively with those threatened by occupational changes, discusses the impact of career crises on workers and offers advice to working adults of all ages.

As the program opens, we hear Professor Lazarus discussing the dramatic changes that have occurred in the life course of work.

LESSON GOALS

By the end of this lesson, you should be prepared to:

1. Discuss psychosocial development during middle age and whether midlife is invariably a time of crisis for men and women.

2. Describe the ways in which family dynamics may change during middle adulthood.

3. Discuss the dynamics of career development during middle adulthood and the ways in which the life course of work has changed.

4. Evaluate the stability of personality throughout adulthood and discuss the normal "unisex" shift that begins late in middle adulthood.

Audio Assignment

Listen to the audio tape that accompanies Lesson 22: "The Life Course of Work."

Write answers to the following questions. You may replay portions of the program if your memory needs to be refreshed. Answer guidelines may be found in the Lesson Guidelines section at the end of this chapter.

1. What historical, economic, and demographic factors have led to the return of large numbers of married women to school and the labor force?

2. In what ways has the life course of work changed in recent generations?

3. What advice is offered in the program for those making occupational choices at ages 20, 40, and 60?

Textbook Assignment

Read Chapter 22: "Middle Adulthood: Psychosocial Development," pages 499–520 in *The Developing Person Through the Life Span, 2/e.*

Write your answers to the following questions. Refer back to the textbook, if necessary. Answer guidelines may be found in the Lesson Guidelines section at the end of this chapter.

1. Discuss the concept of the midlife crisis and the evidence from research on its inevitability in men and women.

2. Describe the ways in which the structure of the "typical" family has changed in recent years.

3. Discuss the concept of the "sandwich generation" and describe how relationships between middle-aged adults and their parents and adult children change at midlife.

4. Discuss the shift in career dynamics that typically occurs during middle age.

5. Describe two aspects of work that are likely to lead to a crisis for the worker at midlife.

6. Discuss whether personality traits remain stable throughout the life span.

7. Discuss the concept of androgyny as it applies to men and women in later life.

Testing Yourself

After you have completed the audio and text review questions, see how well you do on the following quiz. Correct answers, with text and audio references, may be found at the end of this chapter.

1. The return of large numbers of married women to the labor force was largely provoked by:
   a. the Civil Rights Movement of the 1950s and 1960s.
   b. a shortage of unmarried women in the work force following World War II.
   c. efforts of lobbying groups such as the National Organization for Women.
   d. the increase in life expectancy during the past 75 years.

2. Today, more than _____ of women with preschool children are employed.
   a. one-fourth
   b. one-third
   c. one-half
   d. two-thirds

3. A recent national survey of men in their 50s and 60s found that _____% had changed occupations at least once in their lives.
   a. 25
   b. 50
   c. 75
   d. 90

4. Which of the following pieces of advice was offered in the audio program to workers at various stages in their occupational careers?
   a. "Don't expect that your life will be divided into three neat segments corresponding to school, work, and retirement."
   b. "Don't expect that your family will fully understand or support you if you make too dramatic a career change during adulthood."
   c. "Baby boomers will find reduced career opportunities in the years to come, due to shrinking promotional opportunity, pressure from younger workers, and an unwillingness of older workers to retire."
   d. All of the above were offered as advice.

5. According to the experts heard in the audio program, for most professions:
   a. 40 is not too late to begin a new career.
   b. people who do not begin their career until middle age will not have sufficient time to make a significant contribution to their field.
   c. productivity increases directly with the number of years of experience a person has in the field.
   d. all of the above are true.

6. For some people during middle age, dissatisfaction with self, career, or family relationships may develop into a:
   a. personality split.

    b. midlife crisis.
    c. burn-out.
    d. feeling of alienation.

7. The traditional American family is:
    a. an extended family.
    b. increasing in number.
    c. decreasing in number.
    d. a molecular family.

8. When workers feel that they are "cogs in a machine" rather than contributing employees, they are experiencing:
    a. alienation.
    b. burn-out.
    c. the mechanization syndrome.
    d. a midlife crisis.

9. The term "sandwich generation" refers to the feeling of many middle-aged adults that:
    a. they are being pressed on one side by work and on the other by family.
    b. they are being pressed on one side by their adult children and on the other by aging parents.
    c. they are "in the middle," with half of life behind them and half ahead.
    d. all of the above are aspects of their experience.

10. Which of the following was *not* offered in the textbook as an explanation of why some individuals experience a midlife crisis?
    a. Most adults reach a career plateau during midlife.
    b. Changing family dynamics may cause stress during midlife.
    c. People become more aware of their own mortality during midlife.
    d. People become more resistant to change and less tolerant of stress during midlife.

11. Leilani is a teacher who, after 15 years of work, has lost enthusiasm for her career. Teachers such as Leilani, doctors, social workers, and others in the helping professions, are especially susceptible to:
    a. midlife crises.
    b. deindividuation.
    c. alienation.
    d. burn-out.

12. Which of the following personality traits was *not* identified in the textbook as tending to remain stable throughout adulthood?
    a. Neuroticism
    b. Introversion
    c. Openness
    d. Conscientiousness

13. According to Carl Jung's theory of personality:
    a. everyone has both a masculine and a feminine side of their personality.
    b. gender roles become more distinct as people get older.
    c. gender roles are unrelated to social or psychological factors.
    d. gender roles are most distinct during adolescence.

14. Regarding male and female gender roles, David Gutmann has suggested that:
    a. the pressing demands of raising children tend to define gender roles more distinctly.
    b. gender roles become blurred during the normal "unisex of later life."
    c. life experiences are important factors in determining gender roles.
    d. all of the above are true.

15. Which of the following most accurately describes the typical experiences of women adjusting to midlife today?
    a. Because the biological changes of menopause have not changed, adjustment to midlife is no different today than in the past.
    b. Recent cohorts of women have found midlife more difficult to adjust to than previous cohorts.
    c. Compared to previous generations, more women today find midlife a very positive experience characterized by personal growth.
    d. Married women who are mothers show the most successful adjustment to midlife.

LESSON 22 EXERCISE: THE UNISEX OF LATER LIFE

The *Developing Person Through the Life Span, 2/e,* notes that as people get older, both men and women tend to become more **androgynous.** According to personality theorist Carl Jung, every individual has both a masculine and feminine side of personality. During early adulthood, the side that conforms to social expectations is dominant. Then during middle age, men and women become more flexible and feel freer to explore the opposite side of their characters. The sharp gender role distinctions of earlier adulthood break down and each sex moves closer to a middle ground between the traditional gender roles. Many women become more assertive and self-confident. Many men become more considerate, more nurturant, and less competitive as career goals become less important.

Traditional measures of masculinity and femininity are based on the assumption that these traits represent endpoints of a single bipolar dimension that considers the sexes as opposites. Recently, however, several researchers have developed a measure of androgyny based on a reconceptualization of masculinity and femininity as independent dimensions. This test of androgyny—the **PRF ANDRO** scale—contains separate subscales for femininity and masculinity, and is shown on the next page.

To help you to understand better the concept of androgyny, administer the test to two adults of the same sex but in different seasons of life. For example, you might test an adult in his or her 20s, and one in his or her 50s. You might wish to include yourself as one of the subjects. Alternatively, you might ask an older adult to complete the test "as you see yourself now," and "as you were during your early adulthood."

After you have collected your data, determine separate Masculinity and Femininity Scores for each of your respondents by giving them 1 point on each subscale for each answer that agrees with those in the following key. Then complete the questions on the **Exercise Response Sheet** and return only that sheet to your instructor.

Masculinity Key				*Femininity Key*			
2. T	12. T	31. T	47. T	1. T	20. T	37. T	53. T
3. F	15. F	33. T	48. F	5. F	21. T	39. T	55. T
4. T	17. T	34. F	50. T	9. F	22. F	41. T	56. F
6. F	25. T	35. F	52. T	13. T	23. T	43. T	
7. T	26. T	38. F	54. F	14. T	24. F	44. T	
8. T	27. T	40. F		16. F	28. F	45. T	
10. F	29. T	42. T		18. T	32. F	49. T	
11. T	30. T	46. F		19. F	36. T	51. F	

/ / l t / l ı l ı l / ı l /   / / l / ƒ l 7 1 ı ı l / l 7

m9   l ı l /   1 l l 7 l

l l l l 1 l l   l l / l / l / l / / l l 7 l

l / / l / l l

LESSON GUIDELINES

Audio Question Guidelines

1. Until World War II, female employment was largely restricted to young unmarried women. After their marriage or the birth of a child, most women withdrew from the labor force.

   After the war, in the early 1950s, there were approximately four million fewer unmarried women than before the war. Since jobs tend to be gender stratified, employers had no alternative but to begin hiring married women to fill jobs previously held by unmarried women.

   The emergence of continuing education programs for older students occurred in the 1960s.

   These changes have resulted in a resetting of the social clock, such that with each passing decade, the woman returning to the work force is younger because she has been absent from the labor force for a shorter period of time.

   Today, more than half of women with preschool children are employed.

2. One major change is that more and more women today are either returning to the work force or have never left it, despite establishing families and becoming mothers.

   Another change is that the "one job for life" rule no longer holds. In fact, the odds of staying in one occupation for life are getting slimmer all the time. One national survey found that 90% of men in their late 50s and 60s had changed occupations at least once in their lives.

   These changes have led some experts to recommend thinking of work as we do our lives—in terms of seasons.

3. Experts recommend that those just beginning a career take a good look at people who are actually in that occupation and find out what the job is really like.

   Another recommendation is that young people no longer count on a "linear experience" of being educated in their early 20s, working until 65, and then retiring. Rather, experiences should occur in parallel, as people periodically break away from work in order to refresh themselves in school or acquire new skills.

   Experts warn those at midlife—the "outriders of the baby boom generation" —that shrinking promotional opportunity, pressure from the younger generation, and an unwillingness of those who are older to retire, may result in fewer career opportunities in the years ahead.

   For those nearing retirement without any previous commitment to leisure, experts warn that retirement may become a dull, boring trap. Their advice is for middle-aged people to start thinking about part-time work, hobbies, or other ways of maintaining identifiable and satisfying pursuits.

Textbook Question Guidelines

1. The concept of a midlife crisis is supported by the research of Daniel Levinson, who found that between ages 38 and 48, virtually all of his male subjects experienced the need for a reappraisal. Other studies have found that responses to midlife are less predictable and depend on each person's style of coping with problems.

   Many personal changes do occur during middle age, however, including the awareness that one is aging, a new measuring of life in terms of the time remaining, the need to make adjustments in family relationships, and the reaching of a plateau in one's career potential.

   Although it had long been assumed that women inevitably experience a crisis at about the time of **menopause,** there is little evidence that this is true today. Midlife for many women is a time of personal growth and satisfaction. For middle-aged women several decades ago, self-esteem was more likely to be tied to their roles as mothers and wives. This may account for the fact that as these women grew older and those roles diminished, their self-esteem typically fell. Recent surveys no longer find this trend to be true. More women in middle age today are likely to have trained for a career and to pursue work opportunities as sources of self-esteem.

2. In the 1950s, the **nuclear family** (a married couple and their dependent children) made up the majority of households in the United States. Today, following an increase in the number of childless couples and single-parent households, less than one-third of all households are nuclear families.

   The **extended family,** which includes several generations together, is more typical of traditional societies, such as those in China, India, Africa, and Latin America.

3. The relationship between most middle-aged adults and their parents improves with time. This may be the result of a more balanced view of the relationship, and the fact that the elderly today are generally healthy, active, and independent.

   The relationship between middle-aged parents and their adult children also tends to improve throughout middle age, especially if the children weathered adolescence successfully.

   Middle-aged adults are often referred to as the **sandwich generation,** being squeezed by the needs of both the older and younger generations. They are increasingly called upon to provide emotional, financial, and other types of care-giving support to both the older and younger generations. Such care-giving is required much more often of the current generation of middle-aged adults than of previous cohorts. Their parents will probably live longer and their grown children are more likely to be financially dependent than had been the case in earlier generations.

4. During the **maintenance phase** of career development, middle-aged workers generally do not change jobs as often as they once did, having reached the limits of their abilities and/or career opportunities.

   During middle age, workers often redefine their work goals, investing less of their ego in the job, taking a more relaxed approach to work, and being more helpful to others.

Young adults may be motivated by **agency**: the need to be an active agent pushing for personal success. With maturity, the need for **communion** increases and the worker becomes more concerned about the welfare of other workers, the company, and society.

5. **Burn-out** refers to a worker's feelings of loss of enthusiasm for doing the job. Burn-out is particularly prevalent in the helping professions, such as teaching, nursing, medicine, and social work.

   Two remedies for burn-out are reducing expectations about what one can achieve, and sharing responsibilities with other workers.

   Alienation occurs when workers feel that their work is uninteresting and unimportant. Alienation is most likely to occur in large factories in which the individual worker is not recognized as a person, but simply as a means of performing a repetitive task. Alienation is also fostered when there is a hostile relationship between management and workers.

6. The degree to which personality traits remain stable throughout the life span is controversial.

   Many of the various personality traits can be grouped into three basic types, which researchers have found to remain stable as people age: **neuroticism,** characterized by feelings of anxiety, worry, hostility, and depression; **extroversion,** characterized by tendencies to be outgoing, active, and assertive; and **openness,** characterized by receptiveness to new experiences, ideas, and changes in one's life.

   Researchers have also found that **agreeableness** and **conscientiousness** remain stable throughout the life span, and that there is a consistency in traits such as fearfulness, sociability, cheerfulness, anger, and creativity.

7. As people age, undesirable traits become less prominent and both men and women tend to become more **androgynous.** During late middle age, the sharp sex distinctions of earlier adulthood break down, and each sex moves closer to a middle ground between the traditional gender roles. Many women tend to become more assertive and self-confident. Many men become more considerate, more nurturant, and less competitive as their career goals become less important.

   According to Jung, every individual has both a masculine and a feminine side of personality. During early adulthood the side that conforms to social expectations (the social clock) is dominant. During middle age, men and women become more flexible and feel freer to explore the opposite side of their characters.

Answers to Testing Yourself

1. **b.** This shortage caused employers to begin hiring married women. (audio program)

2. **c.** (audio program)

3. **d.** The odds of staying in one occupation for life are getting slimmer all the time. (audio program)

4. **d.** (audio program)

5. **a.** Many experts consider that the freshness a 40-year-old brings to a new career may compensate for the limited number of years he or she has in which to contribute to the profession. (audio program)

6. **b.** For some people, midlife may bring crisis. For others, because of the particular events of their lives, midlife is a very rewarding period. (textbook, p. 500)

7. **c.** The traditional "nuclear family," consisting of a married couple and their dependent children, is on the decline in the United States. (textbook, p. 505)

8. **a.** Alienation is most likely to occur when workers do not see the relationship between their specific labor and the final product of their job. (textbook, p. 513)

9. **b.** In recent years, middle-aged adults have been increasingly called on to provide emotional support, financial assistance, and caregiving to the younger and older generations. (textbook, p. 506)

10. **d.** Each of the other alternatives was offered as an explanation of the midlife crisis. (textbook, pp. 499–504)

11. **d.** After a time, the worker suffering from burn-out becomes frustrated, the result being that his or her high ideals and great expectations turn into bitter disillusionment and emotional exhaustion. (textbook, p. 512)

12. **b.** Neuroticism, openness, and conscientiousness do tend to remain stable throughout adulthood. (textbook, pp. 515–516)

13. **a.** According to Jung, during middle age many individuals are sufficiently well established in their sexual identity that they feel free to explore their opposite side. (textbook, p. 518)

14. **d.** David Gutmann has suggested all of these. (textbook, p. 518)

15. **c.** Today, women in middle age are more likely to have trained for a career, to have been employed for much of their adult lives, to have work opportunities available to them at midlife, and therefore to have a source of self-esteem that women in earlier cohorts did not have. (textbook, p. 503)

References

Gutmann, David L. (1985). The parental imperative revisited: Towards a developmental psychology of later life. *Contributions to Human Development, 14*, 30–60.

Rossi, Alice S. (1980). Life-span theories in women's lives. *Signs, 6*, 4–32.

An eminent psychologist and sociologist, both of whom are heard in the *Seasons of Life* series, discuss issues pertaining to psychosocial development during adulthood.

Late Adulthood: Physical Development

AUDIO PROGRAM: Opening the Biological Clock

ORIENTATION

Lesson 23 of *Seasons of Life* is about the physical changes that occur during late adulthood. As indicated in the textbook, most people's perceptions of these changes are much worse than the reality. **Ageism,** or prejudice against the elderly, fosters stereotypes that are harmful to older adults and serve to isolate the older generation. As the next three lessons will indicate, this season of life, more than any other, has been subject to misinformation and mistaken assumptions.

Is the physical decline that does occur during an individual's 60s, 70s, or 80s an inevitable product of the **biological clock**? In the program "Opening the Biological Clock," Dr. Robert Butler and biologist Richard Adelman point out that many physical changes of late adulthood that once were attributed to aging may be caused by disease, variations in social context, and other factors that are not intrinsic to aging itself.

Another issue addressed in the program and text is why aging, and ultimately death, occur. Since 1900 over 25 years have been added to the **life expectancy** of the average newborn, largely as a result of better health practices and the elimination of **acute diseases** as causes of death. Although the number of years a newborn can look forward to has increased, the **life span**—the biological limit of life—remains fixed at about 100 to 120 years of age.

As science continues to unravel the mysteries of the biological clock, a number of fascinating questions are raised for future research. Why do we die? Can our life span be increased? Is there a master gene that programs when the hour of death will come? What effect will added years of life have on intellectual potential? On the family? On society? What is surprising is the answer of old people themselves to the prospect of living for 130 to 140 years.

LESSON GOALS

By the end of this lesson, you should be prepared to:

1. Give a realistic description of the physical changes that are due to aging itself and those that are a result of external factors such as social context, disease, and others.

2. Describe the **biological clock** as a metaphor for the body's way of timing events such as death, and examine several theories of why aging occurs.

3. Discuss the causes and effects of ageism.

Audio Assignment

Listen to the audio tape that accompanies Lesson 23: "Opening the Biological Clock."

Write answers to the following questions. You may replay portions of the program if you need to refresh your memory. Answer guidelines may be found in the Lesson Guidelines section at the end of this chapter.

1. Define and differentiate **biological clock, life span,** and **life expectancy.** What (if any) changes have occurred in each of these during the past century?

   biological clock

   life span

   life expectancy

2. Explain the "watch in the water" metaphor introduced in the audio program. If the biological clock is the watch, what in the water influences its operation and has an impact on older bodies?

3. How have the age-related incidence rates of **acute** and **chronic disease** changed during the past century? In what ways do these changes complicate efforts to isolate the causes of physical changes associated with aging?

4. Discuss why researchers are "rewriting the book on aging and sexuality," by addressing the following questions.

   a. What are the causes of changes in sexual response that occur with age?

   b. What is the impact of social context on sexuality? How did the policy decision to segregate older men and women living in nursing homes lead to an erroneous conclusion regarding the biology of aging and sexuality?

5. Identify three physical features of growing old that are probably intrinsic to the process of aging.

   a.

   b.

   c.

6. Explain the concept of **programmed death** at the cellular level. What research evidence supports the idea of a genetically based limit to the life of a cell, and the possibility of resetting this limit?

Textbook Assignment

   Read Chapter 23: "Late Adulthood: Physical Development," pages 525–545 in *The Developing Person Through the Life Span, 2/e.*

Write your answers to the following questions. Refer back to the textbook, if necessary. Answer guidelines may be found in the Lesson Guidelines section at the end of this chapter.

1. Define **ageism,** citing sources of ageist attitudes and several reasons why ageist stereotypes should be eliminated.

2. What are the typical age-related changes that occur in each of the following during late adulthood?

   a. brain function

   b. immune system

   c. sense organs and major body systems

3. Briefly explain the following theories of aging.

   a. the wear-and-tear theory

   b. cellular theories

   c. programmed senescence

4. What are some of the characteristics of people who live to a very old age?

Testing Yourself

After you have completed the audio and text review questions, see how well you do on the following quiz. Correct answers, with text and audio references, may be found at the end of this chapter.

1. Which of the following best expresses the concept of the biological clock?
   a. A gene that scientists have discovered that determines life expectancy
   b. A gene that determines life span
   c. A metaphor for the body's many mechanisms of timing
   d. The time frame in which human evolution takes place

2. What is the most probable explanation for the increase in life expectancy that has occurred since 1900?
   a. The biological clock has been reset as a natural result of the evolution of the human species.
   b. Cultural, rather than biological evolution has occurred, in the form of increased knowledge leading to the control of disease.
   c. A decrease in chronic disease has occurred in older people.
   d. The rate of accidents in older people has decreased.

3. Why do researchers find it difficult to determine the physical effects of aging per se?
   a. The biological clock is sensitive to social context.
   b. It is difficult to differentiate the effects of disease from those of aging.
   c. The range of individual differences in behavior increases in older people.
   d. All of the above are reasons that researchers find it difficult to determine the effects of aging per se.

4. What, if any, changes have occurred in life span and life expectancy during the past century?
   a. Life span has increased by 25 years; life expectancy has not changed.
   b. Life span remains the same; life expectancy has increased.
   c. Both life span and life expectancy have increased.
   d. There have been no changes in life span or life expectancy.

5. "Programmed death" refers to the fact that when normal human or animal cells are grown under artificial laboratory conditions:
   a. the cells from species with longer life spans survive longer than cells from species with shorter life spans.
   b. the cells from young individuals survive longer than those from older individuals.
   c. cells will reproduce a finite number of times and then die.
   d. all of the above occur.

6. Prejudice against an age group is called:
   a. generationism.
   b. senescence.
   c. ageism.
   d. gerontophobia.

7. The general thesis of the Hayflick limit is that:
   a. cell division is finite and the total number of cell divisions is roughly related to the age of the cell.
   b. cells multiply infinitely under laboratory conditions regardless of their age.
   c. aging results from impaired function of the immune system.
   d. nutrition determines an organism's longevity.

8. Professionals have helped to reinforce the stereotype of an unhealthy old age because they:

   a. are an unusually prejudiced group.
   b. deal mostly with ill people seeking care, rather than the healthy elderly.
   c. are afraid of growing old, like most people who study aging.
   d. need to justify their professional interest and credentials.

9. Gerontology is the study of:

   a. cellular aging.
   b. senescence.
   c. old age.
   d. diseases of the elderly.

10. The "squaring of the pyramid" refers specifically to:

    a. the decline in birth rate.
    b. the increased longevity of adults.
    c. the changing demography of the U.S. population.
    d. the lowering of the median age of the population from 28 to 23.

11. By the year 2025, it is expected that the proportions of the U.S. population will be evenly divided into thirds according to age:

    a. ages 0–20; 21–40; 41 and over.
    b. ages 0–15; 16–40; 41 and over.
    c. ages 0–29; 30–50; 51 and over.
    d. ages 0–29; 30–59; 60 and over.

12. The biochemical theory of aging based on the fusing of certain molecules with others is called:

    a. cross-linkage.
    b. error catastrophe.
    c. progeria.
    d. the free radical.

13. The theory of aging that compares the human body to a machine, maintaining that as the mileage adds up its parts deteriorate, is the:

    a. error catastrophe theory.
    b. theory of programmed senescence.
    c. wear-and-tear theory.
    d. Hayflick theory.

14. Which of the following is a normal consequence of aging?

    a. The brain becomes smaller in size by late adulthood.
    b. There is reduced production of neurotransmitters in the brain.
    c. Many of the elderly have sleep patterns that would be considered pathological in a younger person.
    d. All of the above are normal consequences of aging.

15. Regions of the world famous for long-lived people share all of the following characteristics *except:*

    a. moderate diet with little consumption of meat and fat.
    b. regular exercise and relaxation.
    c. frequent social interaction with family and friends.
    d. early retirement from work.

LESSON 23 EXERCISE: AGEISM

A central theme of this lesson is that most people's perceptions of aging are inaccurate and reflect **ageist** stereotypes of physical development in late adulthood. As pointed out by experts in the program, these stereotypes stem from our preoccupation with physical decline that is more the result of disease than it is of aging.

Sociologist Bernice Neugarten has drawn a distinction between the **young-old** —the majority of the elderly who are, for the most part, healthy and vigorous— and the **old-old**—those who suffer major physical, mental, or social losses. The text indicates the ironic fact that many professionals who work with the elderly, including those who specialize in **gerontology,** have inadvertently fostered ageism by focusing on the difficulties and declines of the old-old, and by studying the aged residents of nursing homes, who are often infirm.

Think of two elderly adults whom you know: one who fits Neugarten's description of the young-old, and one who fits her description of the old-old. These individuals may be relatives, friends, or even public personalities you have read about. Write a paragraph about each person, briefly describing his or her health, personality, and lifestyle, and explain why you have classified him or her as young-old or old-old. Also indicate the extent to which each person fits, or does not fit, the usual stereotypes of the older adult. Finally, speculate as to why each person developed as he or she did. What losses, for example, might the old-old person have experienced? Return the completed **Exercise Response Sheet** to your instructor.

LESSON GUIDELINES

Audio Question Guidelines

1. The biological clock is a metaphor for the body's many mechanisms of timing.

   Life span refers to the biological limit of life.

   Life expectancy refers to the number of years a typical newborn can expect to live, allowing for the hazards of his or her particular historical time.

   Although life expectancy has increased by 25 years since 1900, the biological clock (and therefore life span) has not changed. Then and now we are timed genetically to have a maximum life span of between 100 and 120 years.

2. In investigating the physical effects of aging, it is important to remember that aging, or the operation of the biological clock (or watch), occurs in a specific historical context in which social, biological, and other types of influences interact with those of aging itself.

   Chronic disease, for example, produces many physical changes (e.g., sexual response, balance) that once were thought to be intrinsic to the process of aging.

   Social context is another variable "in the water" that influences the operation of the biological clock and manifests itself in many behaviors, including sexual response, hormone levels, and even neural connections in the brain.

3. Acute illnesses, such as influenza or the common cold, occur frequently in early life but diminish with age. They have nearly been eliminated as causes of death among older persons in the United States.

   Chronic diseases, such as cancer and heart disease, are uncommon in the young but increase in incidence with age.

   Chronic diseases themselves often produce behavioral changes that are difficult to isolate from the effects of aging per se because the diseases are much more common in older persons. As stated in the program, "You can't say the cause is aging unless you're sure it's not disease."

4. Sexual response does slow down with age, but the cause may be chronic disease rather than aging. Changes in the physiology of circulation and the nervous system that may be related to disease—rather than an inevitable result of aging—may account for this slowing down. These changes may therefore be correctable through the use of certain medications or surgery.

   It was once accepted as fact that the level of the male hormone testosterone declines with age. Dr. Robert Butler has discovered that this "fact" came from studies of older men living separately from women in institutions.

   When these studies have been repeated in recent years, either in community-dwelling men or in institutionalized men who are socialized with women, not only do testosterone levels fail to diminish, but sexual capability is shown not to be lost in normal, healthy elderly men.

5. Three features of the life span that appear intrinsic to the process of aging are a slowing down of behavior (e.g., reflexes, cognitive functioning), an increased variation in behavior, and the longer lives of women.

6. Biologists are convinced that components of our bodies are genetically programmed to die. Biologist Leonard Hayflick discovered that when normal human cells are grown in the laboratory, they will reproduce a finite number of times and then die.

   Cells from older individuals do not survive as long as cells from younger individuals.

   Cells from longer-lived animal species survive longer than cells from species with shorter life spans.

   Introducing genetic material from cells that are nearly dead into younger cells will cause premature death.

   This evidence suggests that at the level of the cell there is some mechanism that is counting and that prescribes the moment of death for the cell.

   Biologists have also found that infecting cells with certain viruses will cause them to divide infinitely, in effect resetting the biological clock so that the cell will not die. Biologists are also acquiring the ability to remove, add, and modify genes—techniques that might affect the biological clock.

Textbook Question Guidelines

1. Ageism refers to the prejudicial stereotypes many persons have about older people.

   Ageism reflects our culture's veneration of youthfulness and the increasing age-segregation of our society.

   Many professionals who work with the elderly have inadvertently fostered ageism by focusing on the declines of old age, and by studying those persons Neugarten refers to as the **old-old**—aged individuals who suffer major physical, mental, or social losses.

   As the shape of the **demographic pyramid** changes to reflect the increased number of elderly persons due to falling birth rates and increased longevity, ageism will be challenged. Today, the fastest growing age group is that of people over age 75.

2. a. During late adulthood the brain loses about 5 percent of its weight, 15 percent of its size, and becomes slower in function. The slowing of neural processes, such as reaction time, may be related to reduced amounts of neurotransmitters and blood flow to the brain.

   b. The immune system, which defends the body against foreign substances, becomes less efficient with age. The thymus gland shrinks, and both **T cells** and **B cells** are less efficient, making diseases such as cancer more common in older adults.

c. Most of the visual and auditory declines of the aged are correctable. Approximately 90 percent of those over age 65 have some visual impairment; approximately one-third experience a hearing impairment that hampers their daily living.

At some point in old age, the slowing of body function and organ reserve necessitates the readjustment of daily routines.

3. a. According to the wear-and-tear theory, body parts simply wear out with normal use and exposure to disease, poor diet, environmental hazards, and other stresses.

b. Cellular theories of aging include the proposal that aging is the result of the accumulation of cellular accidents that occur during cell reproduction **(error catastrophe)**. Another proposes that biochemical **cross-linkages** occur, reducing tissue elasticity. A third theory proposes that oxygen **free radicals** created by the body damage cells and organs, causing diseases such as cancer and atherosclerosis.

c. Many scientists believe in the concept of **programmed senescence**: that the body is genetically programmed to die after a fixed number of years. Some believe that a **genetic clock** actually programs the moment of death, with genes switching off growth processes and switching on aging processes at some predetermined age.

4. Characteristics of long-lived people in parts of the Soviet Union, Pakistan, and Peru include a moderate diet, consisting mostly of fresh vegetables and low levels of fats and meat; continual activity throughout life; an integrated family and community network; daily exercise and relaxation; and a rural environment that minimizes exposure to pollution and maximizes cardiovascular and respiratory fitness.

Answers to Testing Yourself

1. **c.** Although a number of actual mechanisms have been proposed, the "biological clock" is only a metaphor for the body's mechanisms of timing. (audio program)

2. **b.** Life span has not increased, therefore (a) is incorrect. The incidence rates of chronic diseases and accidents in older people have not decreased, therefore (c) and (d) are also incorrect. (audio program)

3. **d.** All of these reasons complicate the effort to determine the physical effects of aging per se. (audio program)

4. **b.** Life expectancy, but not life span, has increased. (audio program)

5. **d.** Programmed death is exemplified by all of these. (audio program)

6. **c.** Generationism and gerontophobia are made-up terms. Senescence refers to the physical decline that occurs with aging. (textbook, p. 526)

7. **a.** This limit according to Hayflick, may determine the life span by not allowing cells to reproduce themselves indefinitely. (audio program; textbook, pp. 539–541)

8. **b.** This natural tendency has resulted in a characterization of older adults that is based on an unrepresentative sample. (audio program; textbook, pp. 526–528)

9. **c.** The study of cellular aging, senescence, and diseases of the elderly are *aspects* of aging that some gerontologists might be interested in. (textbook, p. 527)

10. **c.** The decline in birth rate (a) and increased longevity (b) are factors in the changing demography of the population. (d) is incorrect because the median age is increasing rather than decreasing. (audio program; textbook, p. 528)

11. **d.** Experts warn that new social problems, such as increased expense for medical care and decreased concern for the quality of children's education may result from this trend. (textbook, p. 528)

12. **a.** Error catastrophe is the idea that errors in cell duplication may fatally impair the body's ability to function. Progeria is a genetic disease that causes children to age prematurely. Free radicals are molecules with an unpaired electron that are believed to damage cells, affect organs, and accelerate diseases.(textbook, pp. 537–541)

13. **c.** Cellular theories of aging, such as error catastrophe, the Hayflick theory, and programmed senescence, are more promising than the wear-and-tear theory. (textbook, pp. 536–541)

14. **d.** These are all typical changes that occur with aging. (textbook, pp. 529–533)

15. **d.** Long-lived adults in these regions continue to work throughout life. (textbook, p. 543)

References

Finch, Caleb E., and Schneider, Edward L. (1985). *Handbook of the biology of aging.* New York: Van Nostrand Reinhold

This work is considered a definitive source for research on aging.

Late Adulthood: Cognitive Development

AUDIO PROGRAM: The Trees and the Forest

ORIENTATION

Each season of life has its own way of thinking. Lesson 24 centers on thinking during late adulthood. Chapter 24 of the textbook examines the use of standardized tests and artificial situations in the study of cognition in the elderly, and presents the **information-processing** approach as a useful way to view cognitive development during late adulthood. While intellectual declines are observable for most people in their 60s, and in virtually everyone by age 80, these declines are more apparent in persons performing laboratory tasks than in real life. They are also limited in scope and greatly affected by health, education, and other experiences. In daily life, most older adults are not hampered by changes in their cognitive abilities.

Audio program 24, "The Trees and the Forest," points out that many of the fears of adults as they approach old age are exaggerated. **Alzheimer's disease,** for example, is not nearly as common as public anxiety would suggest. For most people today, late adulthood brings neither a loss nor a disease of memory, but a change of memory. Conscious memory for the short term **(primary memory)** and memory for the very long term **(tertiary memory)** change very little with age. What does decline with age appears to be memory for the immediate term—the kind of memory researchers refer to as **secondary memory.** But even this loss may be more the result of disuse and poor strategies of memorization than aging per se. When older persons are taught better strategies, they often are able to use them effectively.

In this program, psychologist Warner Schaie discusses the cognitive changes that are likely to come with each decade after 60. Marion Perlmutter, another psychologist, reflects on **wisdom,** which may be the special gift of this season.

LESSON GOALS

By the end of this lesson you should be prepared to

1. Explain why different measures of cognition may be needed to assess thinking in older and younger adults.

2. Apply the information-processing approach to cognition in old age.

3. Discuss the potential for new cognitive development, growth, and wisdom during late adulthood.

4. Summarize the causes and effects of the different dementias.

Audio Assignment

Listen to the audio tape that accompanies Lesson 24: "The Trees and the Forest."

Write answers to the following questions. You may replay portions of the program if you need to refresh your memory. Answer guidelines may be found in the Lesson Guidelines section at the end of this chapter.

1. Discuss how the focus of the study of intelligence in older persons has changed in recent years.

2. Describe each of the following memory stages and whether the function of each stage normally changes as an individual ages.

   a. primary memory

   b. secondary memory

   c. tertiary memory

3. Discuss whether the decline in secondary memory is intrinsic to older age or is the result of deficiencies in strategies of **encoding** and **retrieval.**

4. Describe the normal changes in intellectual functioning that occur in most persons during their 60s, 70s, and 80s.

5. Discuss the concept of **wisdom** as introduced in the audio program. What kinds of wisdom may come in later life?

Textbook Assignment

Read Chapter 24: "Late Adulthood: Cognitive Development," pages 547–573 in *The Developing Person Through the Life Span, 2/e.*

Write your answers to the following questions. Refer back to the textbook, if necessary. Answer guidelines may be found in the Lesson Guidelines section at the end of the chapter.

1. In terms of the information-processing model, how are input, storage, and program affected by age?

2. Discuss the relationship of health and education to the age-related decline in cognitive functioning.

3. Explain why the typical memory experiment is often biased against the elderly person.

4. Describe the causes and effects of the following disorders of intellectual functioning.

    a. dementia

  b. Alzheimer's disease

  c. multi-infarct dementia

  d. nonorganic causes of dementia

5. Describe several of the positive cognitive developments that are likely to find expression during late adulthood.

Testing Yourself

After you have completed the audio and text review questions, see how well you do on the following quiz. Correct answers, with text and audio references, may be found at the end of this chapter.

1. The longest-lasting kind of memory is called:
   a. primary memory.
   b. secondary memory.
   c. tertiary memory.
   d. sensory register.

2. Holding material in mind for a minute or two requires which type of memory?
   a. Primary memory
   b. Secondary memory
   c. Tertiary memory
   d. Sensory register

3. New cognitive development in late adulthood is characterized by Professor Perlmutter as:
   a. dementia.
   b. tertiary memory.
   c. wisdom.
   d. encoding.

4. Which type of material would most likely *not* be lost from the memory of a person in his or her 70s?
   a. The names of former business associates

b. Phone numbers of favorite shops
c. Early experiences with the family
d. Dates of appointments

5. Which type of memory typically shows the greatest decline with age?
   a. Primary memory
   b. Secondary memory
   c. Tertiary memory
   d. Sensory register

6. Which of the following is true regarding the rate of decline of various cognitive abilities?
   a. Crystallized intelligence declines more quickly than fluid intelligence.
   b. Fluid intelligence declines more quickly than crystallized intelligence.
   c. Both crystallized and fluid intelligence decline at about the same rate.
   d. There is no decline in either crystallized or fluid intelligence until terminal decline begins.

7. According to the information-processing approach to cognition, older adults show deficits at the *input* stage because:
   a. sensory and perceptual processes decline with age.
   b. short-term memory declines rapidly with age.
   c. long-term memory declines with age.
   d. all of the above changes occur with age.

8. Which of the following describes how cognitive programs tend to change with age?
   a. The memory strategies used by older people to encode information are less efficient than those used by younger people.
   b. Retrieval strategies used by the elderly are less efficient than those used by younger people.
   c. Older adults are less likely to rehearse or to chunk new information.
   d. All of the above changes tend to occur as people get older.

9. The textbook suggests that a more powerful influence on many cognitive measures than aging is the individual's:
   a. health.
   b. education.
   c. sense of humor.
   d. health *and* education.

10. The textbook suggests that the typical laboratory memory experiment puts elderly subjects at a disadvantage because:
    a. older adults generally do not benefit from the repeated practice of such experiments as much as younger adults do.
    b. they are constructed to prevent "priming," which deprives older adults of a retrieval aid they often use in daily life.
    c. older adults tend to become nervous in these test-taking settings.
    d. most such studies focus on practical problem-solving skills.

11. Dementia is a pathological loss of intellectual functioning that is:
    a. always caused by genetic predisposition.
    b. usually progressive and occurs in identifiable stages.
    c. an inevitable consequence of aging.
    d. found only in those over age 60.

12. The form of dementia identified by abnormalities of the cortex, called plaques and tangles, that destroy normal brain functioning is called:
    a. multi-infarct dementia.
    b. Parkinson's disease.
    c. Pick's disease.
    d. Alzheimer's disease.

13. Cognitive malfunctioning may be caused in some older adults by:
    a. temporary obstructions of the brain's blood vessels.
    b. diseases such as Parkinson's disease.
    c. improper use of multiple drugs.
    d. all of the above conditions.

14. Interiority refers to the tendency of many older people to:
    a. forget trivial facts from their lives.
    b. become cynical as they get older.
    c. become introverted and reclusive as they age.
    d. become more reflective and philosophical.

15. Butler calls the attempts of older adults to put their lives into perspective the:
    a. stage of integrity.
    b. life review.
    c. retrospective.
    d. stage of interioration.

LESSON 24 EXERCISE: PERSONAL WISDOM IN OLDER ADULTS

One theme of Lesson 24 is that each season of life has its own way of thinking and its own gift of knowledge. Both the textbook and audio program note that despite the cognitive declines of late adulthood, positive changes occur as well, as the elderly develop new interests, new patterns of thought, and what Professor Marion Perlmutter and others have referred to as **personal wisdom.** Many older adults become more responsive to nature, more appreciative of the arts, more philosophical, and more spiritual. To examine cognitive development in later life, ask an older adult whom you know well to complete the following Life/Values/Goals questionnaire. If you are an older adult, complete the questionnaire yourself. You, or the person you ask to complete the questionnaire, should answer the questions from two life cycle perspectives: as you (or your subject) felt during early or middle adulthood, and as you (or your subject) feel now, during late adulthood. Afterwards, complete the **Exercise Response Sheet** and return it to your instructor.

Source: Bugen, Larry A. (1979). *Death and dying: Theory, research, practice.* Dubuque, Iowa: Brown, p. 457.

LIFE/VALUES/GOALS

As you see your life now, try to answer the following.

1. What three things would your friends say about you and your life if you died today?

2. Given the likelihood that you will not die today, and have some time left to change some things in your life, what three things would you most like to have said about you and your life?

3. If someone were to witness a week of your life, what assumptions would that person make about your values, that is, what matters most to you?

4. What values do you hold that are not evident from the way you live your daily life?

5. What three goals are important to you as you plan your life?

LIFE/VALUES/GOALS

Try to answer the following questions as you might have answered them when you were a younger adult.

1. What three things would your friends say about you and your life if you died today?

2. Given the likelihood that you will not die today, and have some time left to change some things in your life, what three things would you most like to have said about you and your life?

3. If someone were to witness a week of your life, what assumptions would that person make about your values, that is, what matters most to you?

4. What values do you hold that are not evident from the way you live your daily life?

5. What three goals are important to you as you plan your life?

LESSON GUIDELINES

Audio Question Guidelines

1. At one time the study of intelligence in old age was a study of what was believed to be the inevitable loss of function.

   Today researchers realize that each time of life has its own way of thinking, its own gift of knowledge that compensates for other losses.

   For the first time in history, as more people reach their 60s, 70s, and 80s, researchers are testing older persons in an effort to understand how thinking changes as people become older.

2. **a.** Primary memory refers to memory that lasts for a minute or two. It is used to keep information in a person's conscious mind. Primary memory changes very little with age.

   b. Secondary memory refers to the storage of long-term memories that we are not consciously aware of. Secondary memory often becomes less reliable with age.

   c. Tertiary memory refers to memory for things over the long, long term, often involving experiences from early life. Tertiary memory does not decline with age, perhaps because we go over these distinctive memories time and time again.

3. Older persons do not automatically use the best strategies to put information into secondary memory (encoding). This may simply be a result of disuse of such strategies following retirement.

   Older persons can successfully use more efficient encoding and retrieval strategies when they are taught how.

   Examples of encoding and retrieval strategies include the use of imagery, context cues, the formation of rhymes and acronyms (e.g., "H-O-M-E-S," to retrieve the names of the five Great Lakes: Huron, Ontario, Michigan, Erie, Superior).

4. Using a **sequential research** program, in which the same group of people is periodically tested and retested (longitudinal research), and a new sample is added at each testing (cross-sectional research), Professor Schaie has studied people in their 60s, 70s, and 80s.

   The intellectual function of people in their 60s is not significantly different from that of people in their 50s.

   In the decade of the 60s, when most people are entering retirement, the saying "Use it or lose it" becomes an accurate description of intellectual change in older persons. There is often a loss of the selective skills once used in work but no longer needed after a person has retired.

Sometime in the 70s there is an acceleration of physical decline in many people. As a consequence, many people restrict their environment and exposure to new and interesting things, which may accelerate mental decline.

Among persons in their 80s there are wide individual differences in intellectual function, but, as a rule, there are some losses in performance as compared to the 60s and 70s.

5. Professor Perlmutter believes that many older adults possess a form of personal wisdom that results in:

   a. a greater sensitivity to perspective in life, and knowing what is important, and what is not.

   b. a more global and less self-centered view.

   c. a better integration of emotion and cognition.

   d. knowing what to remember and what to forget.

   e. knowing how to compensate for some of the cognitive losses that accompany age.

Textbook Question Guidelines

1. Fluid intelligence, including memory and abstract reasoning, declines faster than crystallized intelligence. This indicates that declines in cognitive function are related to changes in the basic information processes of input, storage, and program. With increasing age the following changes occur:

   a. the sensory receptors become less sensitive.

   b. perceptual processes and the ability to attend selectively to sensory information decrease.

   c. Although some theorists believe that memory decline may be the most significant deficit in the elderly, not all aspects of memory decline equally. As noted in the audio program, secondary memory shows the greatest decline with age.

   d. The mental strategies used by older people to encode, retrieve, and solve problems are less efficient than those used by younger people.

2. Good cognitive functioning and good health are positively correlated.

   People with heart disease or untreated hypertension tend to show reduced intellectual ability, perhaps as a result of deficient delivery of blood to the brain.

   Because it has been estimated that approximately 50 percent of older adults have some sort of cardiovascular problem, it may be that much of the age-related memory decrement is due to this health problem.

   Number of years of education are positively correlated with good cognitive function with aging.

3. The use of meaningless strings of numbers and words makes it particularly difficult for older adults to practice this type of material.

Most memory experiments are constructed to prevent subjects from **priming** (using the retrieval of one item to facilitate that of subsequent items). Older adults are thereby deprived of a memory tool they often use successfully.

Older adults are also less likely to show the same motivation in performing memory tasks that younger people do. When memory and reasoning tests are more interesting and personally relevant, many of the age-related deficits do not appear.

4. **a. Dementia** refers to a pathological loss of intellectual functioning with age. The symptoms include memory loss, rambling conversation, confusion about place and time, and changes in personality.

   b. About 70 percent of the victims of dementia have **Alzheimer's disease.** Alzheimer's disease is associated with abnormalities in the cortex, called plaques and tangles, that destroy normal brain functioning.

   One form of Alzheimer's disease, caused by a dominant gene on chromosome 21, is inherited. Other forms are multifactorial in causation, perhaps caused by a virus or the cumulative effects of environmental toxins.

   c. Multi-infarct dementia, or stroke, is the result of a temporary obstruction of the blood vessels that supply the brain with blood.

   Persons with circulatory problems such as arteriosclerosis and hypertension are at risk for multi-infarct dementia.

   d. The nonorganic causes of dementia include adverse reactions to drugs, the intermixing of prescription drugs, malnutrition (especially B vitamin deficiency), abuse of alcohol, and psychological illnesses such as schizophrenia, depression, and other personality disorders.

5. Erikson finds that older adults are more interested in the arts, children, and the whole of human experience than are younger adults.

   According to Maslow, older adults are more likely to reach self-actualization, with heightened aesthetic, creative, philosophical, and spiritual sensitivity.

   Neugarten refers to the **interiority** of the older adult: an increased tendency toward introspection and reflection.

   Another aspect of the attempt of older adults to put life into perspective is the **life review,** which includes an attempt to connect one's own life with that of future generations, as well as those of the past.

   Older adults are also more likely than younger persons to experience an increase in spirituality.

Answers to Testing Yourself

1. **c.** Tertiary memory is memory for the long, long term. It includes our very first memories of life. (audio program)

2. **a.** Also called short-term memory, primary memory lasts for about a minute in the absence of rehearsal. (audio program; textbook, p. 549)

3. **c.** Perlmutter is just beginning research into wisdom, the ability to see the trees rather than the forest. (audio program)

4. **c.** Early, meaningful life experiences seem to last forever. (audio program)

5. **b.** Secondary memory typically shows the greatest decline with age. (audio program; textbook, p. 549)

6. **b.** Fluid intelligence, including memory and abstract reasoning, declines faster than crystallized intelligence. (textbook, p. 547)

7. **a.** The vision and hearing losses that become increasingly common with age mean that older people cannot even begin to process some information because the stimuli in question remain undetected. (textbook, p. 548)

8. **d.** All of these changes in mental programs tend to coincide with aging. (textbook, pp. 551–552)

9. **d.** On many cognitive measures, an individual's state of health and educational background more accurately predict cognitive function than does age. (textbook, pp. 553–556)

10. **b.** Older adults often rely on priming as a memory aid in their daily lives. (textbook, p. 556)

11. **b.** Dementia occurs in three identifiable, and progressive, stages. (textbook, pp. 561–562)

12. **d.** Alzheimer's disease accounts for approximately 70 percent of all cases of dementia. (textbook, pp. 562–564)

13. **d.** It is not uncommon for the elderly to be thought to be suffering from some form of brain disease when, in fact, their behavior is caused by one of these other factors. (textbook, pp. 565–568)

14. **d.** This philosophical turn of the mind represents an attempt to put life into perspective. (textbook, p. 570)

15. **b.** In recalling and recounting life's experiences, one connects with those who have come before, and those who will follow. (textbook, p. 570)

References

Perlmutter, M., and List, J. A. (1982). Learning in later adulthood. In T. M. Field, A. Huston, H. C. Quay, L. Troll, and G. E. Finley (eds.), *Review of human development*. New York: Wiley.

Psychologist Marion Perlmutter, who is introduced in the audio program, offers a lucid critique of the laboratory research findings showing that older adults learn less well than younger adults.

Averyt, A., Furst, E., and Hummel, D.D. (1987). *Successful aging: A source book for older people and their families*. New York: Ballantine.

Written in a sprightly but matter-of-fact tone, this book is a useful resource for the older adult. In addition to dispelling a number of myths of aging, it discusses such major topics as Alzheimer's disease, alcoholism, divorce, and death.

Late Adulthood: Psychosocial Development

AUDIO PROGRAM: Three Grandparents

ORIENTATION

How is psychosocial development affected by old age? In discussing three major theories—**disengagement theory, activity theory,** and **continuity theory**—the textbook notes that the elderly are more diverse in lifestyle, income, and personality than people in any other season of life. This great diversity is not adequately explained by any single theory. The chapter also discusses pressures that come with old age, including retirement, changing relationships with children and grandchildren, decreased income, failing health, widowhood, and the imminence of death.

In the audio program "Three Grandparents," we meet three very different grandparents whose stories address three questions: What is the right age to become a grandparent? How has the dramatic rise in divorce rates affected grandparents? What are the effects of this century's great increase in length of life on the experience of being a grandparent?

Most people become grandparents during their 40s or 50s. Developmentalist Linda Burton has found that **on-time** grandparents are happier and better prepared than those who have the role of grandparent thrust on them early, in their 20s or 30s.

Sociologist Andrew Cherlin notes that in the wake of divorces grandparental ties are normally weakened on the father's side and strengthened on the mother's side. He points out that increased longevity means that many grandparents can now expect to spend half of their lives with their grandchildren and that many will become great-grandparents. Although grandparenthood has changed, it still remains a central role in the family structure.

As the program opens, we hear the voices of three grandparents, each beginning his or her own unique story.

LESSON GOALS

By the end of this lesson, you should be prepared to:

1. Discuss the impact of grandparenthood on the individual and the family.

2. Explain how being a grandparent has changed during the past century.

3. Discuss the psychosocial development of older persons from a variety of theoretical perspectives, while recognizing the diversity of their individual experiences.

4. Describe the special problems that face some older adults in contemporary America, including adjustment to retirement, frail health, poverty, changing relationships, and widowhood.

Audio Assignment

Listen to the audio tape that accompanies Lesson 25: "Three Grandparents."
    Write answers to the following questions. You may replay portions of the program if you need to refresh your memory. Answer guidelines may be found in the Lesson Guidelines section at the end of this chapter.

1. Discuss the impact of being an "on-time" or "off-time" grandparent.

2. Discuss how grandparents are affected by the divorce of their children. What effect does divorce usually have on grandparental ties on the mother's side? On the father's side?

3. Describe the three phases of grandparenting. How does the grandparent's relationship to the grandchildren change during these phases?

4. Compare and contrast the typical role of today's grandparent with that of grandparents at the turn of the century. How was the role of grandparents then different from that of grandparents now?

    c. It is about the same as that between grandparent and grandchild.

    d. It is often much more argumentative since the historical gap between them reflects very different cultural values.

5. Sociologist Andrew Cherlin suggests that grandparents are in the difficult position of having to balance two different desires with respect to their role in the family:

    a. the desire to maintain equally close relationships with their children and their grandchildren.

    b. the desire to make sure the grandchildren are brought up "in the right way" and yet do so without insulting their children by questioning their competence as parents.

    c. to feel secure that they will be cared for in their old age and yet not be a burden to their children or grandchildren.

    d. the desire to be autonomous and yet to retain an important role in the family.

6. The idea that people and society mutually withdraw from each other during late adulthood is a basic tenet of:

    a. continuity theory.

    b. discontinuity theory.

    c. disengagement theory.

    d. activity theory.

7. Researchers who believe that elderly persons are happier if they maintain social contacts rather than withdrawing from society would be called:

    a. continuity theorists.

    b. discontinuity theorists.

    c. disengagement theorists.

    d. activity theorists.

8. Which of the following statements most accurately describes psychosocial development in late adulthood?

    a. Several leading gerontologists believe that people become more alike as they get older.

    b. Older adults generally fit into one of two distinct personality types.

    c. Many gerontologists believe that the diversity of personalities and patterns is greater among the elderly than any other age group.

    d. Few changes in psychosocial development occur after middle adulthood.

9. People who feel unwanted and discarded upon retiring from full-time work:

    a. are usually never able to adjust.

    b. usually find that such feelings are temporary.

    c. should never retire.

    d. should seek immediate psychological counseling.

10. A person's family members, friends, and acquaintances who move through life with him or her are called his or her:

    a. social clock.

    b. extended family.

    c. social convoy.

    d. social cohort.

11. Marital satisfaction among older adults tends to be:

    a. lower than that among younger adults.

    b. about the same as that among younger adults.

    c. higher than that among younger adults.

    d. higher than among younger adults until retirement.

12. The number of frail elderly is currently _____ than the number who are active, financially stable, and capable; the frail elderly are _____ in number.

    a. greater . . . decreasing

    b. less . . . increasing

    c. greater . . . increasing

    d. less . . . decreasing

13. A problem with many nursing homes is that they:

    a. focus more on physical needs than psychological needs.

    b. are often understaffed.

    c. reinforce dependence in patients.

    d. show all of the above problems.

14. Jack, who is 73, looks back on his life with a sense of pride and contentment; Eleanor feels unhappy with her life and that it is "too late to start over." In Erikson's terminology Jack is experiencing _____, while Eleanor is experiencing _____.

    a. generativity . . . stagnation.

    b. identity . . . emptiness.

    c. integrity . . . despair.

    d. completion . . . termination.

15. Generally speaking, the relationship between older adults and their grown children is:

    a. viewed as loving and close.

    b. viewed as distant and stormy.

    c. more important than interaction with friends of their own age.

    d. less important than interaction with their grandchildren.

LESSON 25 EXERCISE: GRANDPARENTS

A number of issues pertaining to grandparents are raised in the textbook and audio program. These include the changing role of grandparents due to increased life expectancy, greater geographical mobility of offspring, greater financial independence, rising divorce rates, and the trend toward egalitarian relationships with grandchildren.

To help you apply the information in this lesson to your own life, reflect on the changing role of grandparents by writing brief answers to the questions on the **Exercise Response Sheet**. Send the completed sheet to your instructor.

LESSON GUIDELINES

Audio Question Guidelines

1. Developmentalist Linda Burton has found that grandmothers who entered the role early (in their 20s or 30s) were not as happy with their new role as were **on-time** grandmothers (those who became grandmothers in their 40s or 50s). The early transition to grandmotherhood seemed to throw their life course out of synchronization. Their rejection of the grandmother role amounts to their putting up a "speed bump" in their developmental cycle, i.e., saying in effect, "Stop, you're trying to push me into a middle-age stage of development."

   On-time grandmothers were much happier with the transition to the grand-mother role because it came at a time in their lives when they expected it and when they were prepared for it.

2. The 1960s and 1970s saw a dramatic rise in divorce rates. Today one of every two marriages is likely to end in divorce. It is therefore not surprising that divorce is having a tremendous impact on grandparents, often weakening their ties to their grandchildren.

   Sociologist Andrew Cherlin has found that whether a daughter or son is getting the divorce often makes a big difference. In most cases of divorce, mothers keep custody of the children. For a grandparent, that means that if your daughter is getting divorced, you are likely to have a closer relationship with the grandchildren because you are more likely to be called in to help. If it is a son getting divorced, it often is more difficult for the grandparent to maintain a relationship with the grandchildren.

3. Phase one of grandparenting begins with the birth of the grandchild. This phase is the best, according to most grandparents. Sociologist Cherlin has found that grandparental ties and emotional investment are highest during this phase.

   Phase two occurs during the grandchild's adolescence. During this phase the grandchildren need to establish independence from their parents and grand-parents. The relationship between grandparents and grandchildren becomes more distant.

   Phase three begins when the grandchild becomes an adult. During this phase the relationship often again becomes close. In many cases the usual roles of grandparent and grandchild start to be reversed: it is the grandchild who is likely to provide more in the way of support to his or her elderly grand-parent.

4. In this century grandparenthood has become a much lengthier period of life and more distinct from the period of parenthood.

Because of the increase in life span, most people can now expect to spend an entire career—perhaps even half their lives—as grandparents.

Increased longevity has made great-grandparenting much more common today than it was at the turn of the century.

Unlike grandparents, great-grandparents almost never become involved in a "hands-on" way with their great-grandchildren. This may be due to the fact that there are too many layers of family (parents and grandparents) insulating them from their great-grandchildren. In addition, most great-grandparents are much older and, although they love their great-grandchildren, they are not as interested in, or capable of, helping out as they were with their grandchildren.

5. Although grandparents today range in age from their 20s to their 60s, and come in infinite varieties, grandparents share several common features.

For one, they are a kind of family insurance policy, standing in the background ready to help out in the case of a family crisis, such as parental illness, divorce, or unemployment.

A second common feature is the norm of noninterference. Grandparents are supposed to leave parenting to the parents; as a result, they must balance two very different desires: retaining their own autonomy, while maintaining a strong role in the family.

Andrew Cherlin believes that grandparents' role as symbols of family continuity and love may actually increase in the future. This is due to the fact that as family size declines there are fewer grandchildren for grandparents to spend time with, giving the average grandparent more resources and more time to devote to a grandchild.

Textbook Question Guidelines

1. a. The most controversial theory of adult psychosocial development, disengagement theory, maintains that during late adulthood the individual and society mutually withdraw from each other. The elderly adopt a passive style and their social circle becomes increasingly smaller as they relinquish many of their previous roles.

b. According to activity theory, disengagement often is not voluntary, and is not a sign of normal aging. Furthermore disengagement in one area may lead to reengagement in others. Activity theorists maintain that the more activities older adults engage in, the happier they are.

c. Another controversial issue is the question of whether psychosocial development is gradual and steady (continuity), or occurs in abrupt transitions (discontinuity). Because retirement, failing health, and the death of a spouse are likely to occur in late adulthood, discontinuity is inevitable. Continuity in friends, interests, and personality is, however, also characteristic of this stage.

2. In terms of history, retirement is a recent phenomenon. Earlier and better-financed voluntary retirement frequently allows adults to redirect their energies into self-chosen activities.

   For some, retirement is extremely difficult to adjust to. Most, however, eventually adjust and enjoy their retirement.

   Some gerontologists have suggested that workers should have the option of gradually retiring, rather than being subjected to the usual abrupt transition.

   The need for achievement and self-esteem continues into old age even though what once were the two main sources of achievement—employment and child-rearing—are completed.

3. The need for affiliation in late adulthood is even stronger than the need for achievement. Older adults' satisfaction with their lives is highly correlated with the quality and quantity of their contact with their **social convoy.** This explains why many remain in established neighborhoods consisting of friends, churches, cultural sources, and social clubs frequented by friends of their own background.

4. Because of lower death and divorce rates in this cohort, there are more married older adults in the current generation than there were in the past.

   Husbands and wives tend to be happier with their marriages in old age than at any other time in their marriage.

   Older marriages are less likely to be characterized by serious disagreements.

5. Approximately 47 percent of adults over 65 live alone. The divorced older adult generally has adjusted well to the changes brought by divorce.

   Death of a spouse is one of life's most serious stresses, involving loss of a friend and lover, lower income, less social status, and a broken social circle.

   Remarriage during late adulthood tends to be happier than remarriage earlier.

   The **frail elderly** are a growing minority. The elderly poor are more likely to be female, widowed, and black or Hispanic.

   A chief consequence of poverty in old age is inadequate health care. Many older adults continue to be cared for by family members. The stress of such care may lead to mistreatment, or **elder abuse.**

   Valuing their independence, many of the elderly feel that nursing homes should be avoided at all costs.

   Nursing homes often place priority on physical needs, to the neglect of psychological needs such as independence, activity, and self-esteem.

Answers to Testing Yourself

1. **c.** Most people become grandparents during their 40s or 50s. This figure has not changed significantly during the past century. (audio program)

2. **b.** Early grandmothers often are not as happy with their roles as are "on-time" grandmothers. (audio program)

3. **a.** Grandparental ties usually strengthen on the mother's side following divorce. This is due to the fact that in most cases of divorce the mother retains custody of the children. (audio program)

4. **b.** Great-grandparental ties are usually weaker than are grandparental ties. (audio program)

5. **d.** Grandparents desire autonomy, yet wish to retain an important role in the family. (audio program)

6. **c.** Disengagement theory has provoked protest from many gerontologists, who disagree with the implication that the elderly disengage, or that when they do, they do so willingly. (textbook, pp. 575–576)

7. **d.** According to activity theory, if disengagement does occur, it should be looked on as a sign that something is wrong, and not as a normal part of aging. (textbook, pp. 576–577)

8. **c.** In fact, given the diversity of health and psychosocial experiences, age as an index to developmental patterns becomes increasingly less relevant as one becomes older. (textbook, p. 578)

9. **b.** Although for most people retirement is "a major economic, social, and psychological event," most older adults quickly find more than enough to do with their new-found freedom. (textbook, pp. 582–582)

10. **c.** One's social convoy continues to be important in late adulthood as a source of happiness and as a buffer against adversity. (textbook, pp. 588–589)

11. **c.** Older marriages are less likely to be characterized by serious disagreements: apparently the conflicts of the younger years have disappeared, diminished, or no longer seem worth disputing. (textbook, p. 591)

12. **b.** The number of frail elderly is growing because of increased life expectancy, improved life-saving techniques, and inadequate health care for many who cannot afford private care. (textbook, pp. 595–596)

13. **d.** For all of these reasons, many older adults feel that nursing homes should be avoided at all costs. (textbook, pp. 600–602)

14. **c.** Recent research has found that the process of achieving integrity is much like that of achieving identity in adolescence. (textbook, p. 603)

15. **a.** Although a "generation gap" between the elderly and their children and grandchildren is common, most of the elderly view these relationships as loving and close. (textbook, p. 584)

References

Cherlin, A., and Furstenberg, F. (1986). *The new american grandparent*. New York: Basic Books.

Professor Andrew Cherlin, who is heard on the audio program, discusses the changing role of grandparents in contemporary society.

Death and Dying

AUDIO PROGRAM: Of Seasons and Survivors

ORIENTATION

The final lesson of *Seasons of Life* is concerned with death and dying. Depending on a person's age, experiences, beliefs, and historical and cultural context, death can have many different meanings. Chapter 26 of *The Developing Person Through the Life Span, 2/e,* explores these meanings, including the recent shift toward a more accepting view of death, due largely to the pivotal work of Elizabeth Kübler-Ross, whose findings, based on interviews with terminally ill patients, helped make professionals and the general public more aware of the needs of the dying.

The textbook also discusses the concept of a "good death," which most agree is a death that comes swiftly and with dignity. It describes the **hospice** as an alternative to the dehumanization of hospital deaths and presents some of the controversy surrounding **euthanasia.**

The chapter concludes with a discussion of **bereavement, mourning,** and the many factors that influence how people think about death. Consistent with our culture's tendency to conceal death by placing the dying in institutions, the process of grieving has recently been denied, which can have a crippling effect on the lives of the bereaved.

Audio program 26, "Of Seasons and Survivors," examines how the ending of one person's life affects the continuing stories of the family members who survive. Through the stories of two very different deaths, the listener discovers that losing a loved one "in season," at the end of a long life, has a very different effect from losing a loved one "out of season," in the prime of life. Expert commentary is provided by psychologist Camille Wortman, who has studied the grieving process extensively.

As the program opens, we hear the voice of a father describing the tragic death of his 30-year-old son.

LESSON GOALS

By the end of this lesson, you should be prepared to:

1. Discuss our culture's attitudes toward death and dying and how these affect the process of adjusting to bereavement.

2. Identify factors that can make dying an easier process and bereavement less traumatic for survivors.

3. Discuss death as a way of giving meaning to life and as a necessary part of the developmental process.

4. Contrast the impact of an "in-season" death and an "out-of-season" death on surviving family members.

Audio Assignment

Listen to the audio tape that accompanies Lesson 26: "Of Seasons and Survivors."

Write answers to the following questions. You may replay portions of the program if you need to refresh your memory. Answer guidelines may be found in the Lesson Guidelines section at the end of this chapter.

1. Identify two misunderstandings the general public has about the grieving process.

2. Describe two ways in which well-intentioned persons are often unhelpful to those who are grieving.

3. Explain how people *can* be helpful to those who have lost a loved one and identify other factors that facilitate the recovery process.

4. Contrast the impact of "in-season" and "out-of-season" deaths on surviving family members.

Textbook Assignment

Read Chapter 26: "Death and Dying," pages 607–620 in *The Developing Person Through the Life Span, 2/e.*

Write your answers to the following questions. Refer back to the textbook, if necessary. Answer guidelines may be found in the Lesson Guidelines section at the end of this chapter.

1. Cite several ways in which our culture has made it more difficult for individuals to cope with death and dying.

2. Discuss how people's views of death change over the life span.

3. Describe the five emotional stages a dying person goes through in dealing with imminent death, as described by Elizabeth Kübler-Ross.

4. Explain what is meant by a "good death," and list reasons for supporting and reasons for opposing the controversial practice of passive euthanasia.

5. Describe the hospice approach to the treatment of dying patients.

6. Outline the four stages of mourning.

Testing Yourself

After you have completed the audio and text review questions, see how well you do on the following quiz. Correct answers, with text and audio references, may be found at the end of this chapter.

1. In her research with grieving families, Professor Wortman has found that:
   a. most people lose their attachment to the lost loved one within a year of the death.
   b. people often remain attached to a lost loved one for many years.
   c. people who remain attached to a lost loved one need professional counseling.
   d. women are better than men at adjusting to the loss of a loved one.

2. Which of the following statements would probably be the most helpful to a grieving person?

   a. "It must have been his or her time to die."
   b. "Why don't you get out more and get back into the swing of things?"
   c. "It must have been God's will that she or he was taken from you."
   d. "If you need someone to talk to, call me at any time."

3. Professor Wortman found that the single most helpful element in a person's recovery from the loss of a loved one was:

   a. the person's getting back to their normal routine.
   b. the person's belief in a religious faith.
   c. having an understanding relative.
   d. having a friend who had experienced a similar loss.

4. Which of the following is true concerning the timing of death?

   a. Today, death is more likely than ever to come in any season of life.
   b. Today, death is more likely to come at the end of a long life.
   c. The timing of death has not changed significantly over the course of human history.
   d. Death is less predictable today than ever before.

5. In helping a person cope with the loss of a loved one, it is *not* a good idea to:

   a. rush people through their grief.
   b. provide a philosophical perspective on the death.
   c. encourage people to forget and get on with life.
   d. do any of the above.

6. Throughout most of history, death has been viewed as:

   a. an event that should be handled outside of the home.
   b. an event to be kept from younger children.
   c. an event to be forgotten as quickly as possible.
   d. a familiar and accepted part of the life cycle.

7. During the twentieth-century in Western countries, death came to be:

   a. withdrawn from everyday life.
   b. more accepted as part of everyday life than in earlier times.
   c. concealed by everyone except those in the medical profession.
   d. discussed more openly than at any other time in history.

8. A 5-year-old child is *unlikely* to believe that:

   a. death happens only to those who want to die or are evil.
   b. death is a universal event.
   c. death is a temporary state.
   d. dead things can be revived.

9. By about age _____ children begin to recognize that death is an inevitable, biological phenomenon.

   a. 7
   b. 9
   c. 11
   d. 13

10. Having recently been told that he will soon die, Jack refused to make any plans or take interest in his treatment. According to Kübler-Ross, Jack is probably in the stage of:

   a. denial.
   b. anger.
   c. bargaining.
   d. depression.
   e. acceptance.

11. Researchers other than Kübler-Ross who have interviewed the dying have found that:

   a. the five emotional stages of dying usually occur in the sequence Kübler-Ross specified.
   b. none of the stages proposed by Kübler-Ross commonly occur in those who are dying.
   c. Kübler-Ross's stages are not applicable to everyone or to every context.
   d. the stages occur only in those who experience a "good death."

12. Which of the following is an example of passive euthanasia?

   a. Giving a dying patient a lethal injection
   b. Disconnecting the patient from a life-sustaining piece of equipment
   c. Withholding a special drug needed to sustain the patient's life
   d. Allowing the dying patient to commit suicide

13. According to the textbook, dying a "good death":

   a. was more likely in earlier centuries when little could be done medically to postpone death.
   b. was less likely in earlier centuries when little could be done medically to postpone death.
   c. often means prolonging life through artificial life-support systems until surviving family members are convinced that nothing else can be done to save their loved one.
   d. usually means different things to the dying person and family members.

14. Historically, the practice of mourning:

   a. was considered self-indulgent and was discouraged.
   b. focused on rituals that helped the grieving person avoid confrontation with their true feelings.
   c. helped those who had lost loved ones to express their grief.
   d. was based on the practice of mummification.

15. In confronting the deaths of friends and family members, elderly persons:
    a. are more likely than younger adults to withdraw from society.
    b. are more likely than younger adults to engage in practices that deny death.
    c. are usually less accepting of death than younger adults.
    d. are often subject to the stress of bereavement overload.

LESSON 26 EXERCISE: COPING WITH DEATH AND DYING

A central theme of this lesson is that death has many meanings, depending on a person's age, experience, beliefs, and his or her historical and cultural context. A number of issues are discussed, including our culture's tendency to institutionalize and deny death, the decline of "good deaths" as medical technology has provided new ways of prolonging life, developmental shifts in the way people view death, euthanasia, and the pros and cons of the hospice as an alternative to hospitals for those who are dying.

To stimulate your thinking about these issues, complete the questions on the **Exercise Response Sheet**. If you would like to ask these questions of someone other than yourself, feel free to do so. If you need more space for the answers, you may use additional sheets of paper. Return the completed response sheet to your instructor.

LESSON GUIDELINES

Audio Question Guidelines

1. In her research on the process of grieving, Professor Camille Wortman has found that the general public has many misconceptions about grief. One is the belief that those who are bereaved are eventually able to break their attachment to the lost loved one. In reality, it is very common for the bereaved to remain attached to a lost loved one for a long time.

   Another misunderstanding is that after a year or two those who have lost a loved one will recover and get back to their normal routine. Professor Wortman has found that the vast majority of people who have lost a loved one experience permanent changes in their lives as a result.

2. Well-intentioned people often attempt to relate to people who are grieving in ways that are not helpful. One thing that usually is not helpful is to provide a philosophical perspective on the event, such as the comment that "Well, it was her time to die."

   Another unhelpful thing is to try and hurry people through their grief by encouraging them to "get back into the swing of things."

3. Professor Wortman suggests that bereaved persons are comforted by social support, such as a friend who listens, sympathizes, and does not ignore the real pain and complicated emotions that accompany mourning.

   Those who would comfort the bereaved should also realize that bereavement is likely to be a demanding process that may last for months or even years.

   Professor Wortman also found that recovery from the loss of a loved one was facilitated by some form of religious faith.

4. Every life, short or long, leaves a legacy and lasting effect on surviving family members. The impact of losing a loved one in the prime of his or her life is different from that of losing a loved one at the end of a long life. The sudden death of a person who is not "supposed" to die, such as the young man in the audio program, is usually the most difficult to bear. Surviving members of the family are often tormented by conflicting emotions of guilt, denial, anger, and sorrow.

   Death is somewhat easier to cope with when it is expected. Losing someone unexpectedly does not allow family members to come together with the dying person and share their affection for each other. Having time to anticipate and prepare for the death does not necessarily reduce the pain of loss, but it can reduce the conflicting emotions associated with it.

Textbook Question Guidelines

1. With advancements in medical technology in the twentieth century, death has been removed from the daily life experiences of most people.

   The denial of death has come to permeate the medical profession, even to the point that the dying experience a "social death" because they are avoided by doctors and nurses.

   Consistent with the Western custom of denying death, the practice of

mourning has declined. In earlier times, the bereaved were encouraged to express grief openly and fully, often following a prescribed ritual and schedule.

When grief is denied, it may cripple a person's life, even leading to suicide or practices such as **mummification,** in which the bereaved individual leaves intact the belongings of the dead.

2. Children between 3 and 5 deny the permanence of death. Those between 6 and 9 believe in the existence of death but think that only certain people, such as those who are bad, die. Children older than 9 recognize death as a universal, inevitable biological event.

   Children's thinking about death is also related to their stages of cognitive development, with preoperational children having a more egocentric concept of death than children who are capable of concrete or formal operational thought.

   Adults' anxiety about death depends on the particular circumstances of their lives. Unmarried men, for example, are more anxious about death than married men. Married men with children are less anxious than married men without children. Among women, exactly the opposite is true: mothers have the highest level of anxiety about death.

3. Kübler-Ross's stages of dying begin with **denial,** during which the patient refuses to believe that he or she will die.

   Denial is followed by **anger** and then **bargaining,** in which the patient tries to negotiate an alternative, promising God or fate, for example, that he or she will live a better life if given the chance.

   The fourth stage of **depression** causes the person to mourn his or her own death and be unwilling to make any plans or take an interest in treatment. Finally, **acceptance** may occur, as the person understands that death is the last stage of this life and, perhaps, the beginning of the next.

4. Most people agree that a "good death" is one that occurs swiftly and allows the individual to die with dignity while surrounded by loved ones.

   In earlier times, when little could be done medically to postpone death, "good" deaths were more likely than they are today, when many who die in hospitals are likely to die in pain or in a state of semiconsciousness.

   Unlike **active euthanasia** (mercy-killing), **passive euthanasia** involves inaction that allows a person to die. Many health care professionals would like to see passive euthanasia become common practice in order to improve patients' chances of dying a good death. Disagreements about the quality of life, the ability to predict the exact course of illness, and when death will occur, however, make euthanasia a controversial topic.

5. In order to help more of the terminally ill die a good death, Cecily Saunders opened the first hospice in London during the 1950s.

   Hospices provide the dying with skilled medical care, but not artificial life-support systems. Often held in a home or dormitory setting, the hospice setting respects the patients' dignity, allowing them to wear their own clothes, have visitors at any time, and move about with no restrictions.

The hospice concept has its critics, however, and raises many legal questions. In addition, it is only available to a minority of dying people who are diagnosed as terminally ill and with no chance of recovering.

6. The first phase of mourning is characterized by **shock** and numbness. Some aggrieved people react in a calm, rational fashion, while others seem dazed.

During the second phase the mourner experiences an intense longing to be with the lost person.

The third phase is one of **depression** and despair, during which the grieving person may experience irrational anger and confused thinking.

In the final, **recovery** stage, the death is put into perspective.

Anniversary reactions—often occurring on holidays, birthdays, or the anniversary of the death—may interrupt recovery and trigger a new period of mourning.

How long it takes to recover from the death of a loved one depends on how well the particular culture facilitates mourning by providing meaningful customs, emotional support, and practical help to the mourner.

Answers to Testing Yourself

1. **b.** It is a common misconception that the bereaved are eventually able to break their attachments to loved ones who have died. (audio program)

2. **d.** Professor Wortman's research has consistently found that the social support of a friend who is available to listen is an important aspect of the recovery process. (audio program)

3. **b.** Having (c) an understanding relative and/or (d) a friend who has experienced a similar loss are/is also helpful to recovery, but religious faith seems the most beneficial of these factors. (audio program)

4. **b.** Death is more likely to come "in season" today as a result of improved health practices and control of disease. (audio program)

5. **d.** Altering the natural process of grieving by (a) rushing it, (b) attempting to rationalize it or (c) denying it can have a crippling effect on the bereaved. (audio program)

6. **d.** Only in the Western culture of the twentieth century have death and mourning been denied. Previously, death was a familiar, and natural, occurrence in everyday life. (textbook, p. 607)

7. **a.** In the twentieth century more and more people are dying alone in hospitals rather than at home among their families. (textbook, pp. 607–608)

8. **b.** Five-year-olds are likely to believe that death is temporary and that it happens only to certain people. (textbook, p. 609)

9. **b.** At about age 9 children begin to recognize death as a universal and inevitable phenomenon. (textbook, p. 609)

10. **d.** Jack's behavior is typical of Kübler-Ross's stage of depression, in which the person mourns his or her own impending death. (textbook, p. 611)

11. **c.** Researchers typically find that denial, anger, bargaining, and depression appear and reappear during the dying process, depending on the context of the death. (textbook, pp. 611–612)

12. **c.** Unlike active euthanasia, in which someone must perform an act that kills the individual (a and b), passive euthanasia involves inaction that allows someone to die. Answer d is incorrect because in a suicide the person causes his or her own death. (textbook, p. 613)

13. **a.** Today, because modern medicine can frequently sustain life beyond its "time," many people die in pain or semiconsciousness. (textbook, p. 612)

14. **c.** In earlier times, the bereaved were encouraged to express their emotions openly and fully. This usually facilitated the recovery process. (textbook, p. 616)

15. **d.** Although the elderly are generally accepting of death, they are vulnerable to bereavement overload, since a second or third death can restart the mourning process before their mourning of prior deaths has been completed. (textbook, p. 617)

References

Kastenbaum, R. (1985). Dying and death: A life-span approach. In J. Birren and K.W. Schaie (Eds.), *Handbook of the psychology of aging*. New York: Van Nostrand Reinhold.

This excellent chapter provides a comprehensive overview of attitudes toward death across the life span.

The Television Term Project

OVERVIEW

The five television programs of *Seasons of Life,* hosted by David Hartman, present the stories of people at all stages of life. The programs may be viewed any time during the semester and perhaps more than once. After you watch each one, respond to the questions designated. These questions will help you integrate the television programs with other aspects of the course.

If possible, watch the television programs with other people—students in the course, family, or friends. Feel free to discuss the questions with them. Then write your own answers.

In some cases you will have to choose between questions for "younger" and "older" students. No matter what your age, pick the question better suited to you. Once you have answered all twenty-five questions, send the entire set of your answers to your instructor.

PROGRAM ONE: INFANCY AND EARLY CHILDHOOD
(Conception to age 6)

Program One follows the biological, social, and psychological clocks through the first six years of life. It explores the development of attachment, autonomy, gender identity, autobiographical memory, and the sense of self. It addresses the controversial issues of day care and expert "advice" on how to raise children. The program also discusses the dramatic changes that have occurred in the seasons of life, and introduces some of the families and experts who will appear throughout the series.

STORIES

- The Kennedy family of Butler, Pennsylvania, takes great pride in the 50 years and four generations they have worked the family dairy farm. At the head of the family are Martha, 72, and Francis, 69. Grandson Jeffrey and his wife Janice, both 22, have just become the proud parents of the newest Kennedy, Justin, whose birth brings back memories for each member of the family. Martha and Francis remember the birth of Jim, Justin's grandfather, and ponder how different will be the world in which Justin grows up. Jim and wife Rita, both 45, wonder if Jeffrey and Janice are too young to be having a baby. Says Rita, "When I see people today with babies I think, Oh boy! Are they young! But we were at that stage one time too. I have to remember that."

- A day care center in Pittsburgh is the setting for several vignettes about the effects of multiple care-givers on children. One such child is 5-week-old Grant Templin, whose mother Diane is returning to her full-time job. "It's very difficult for me as a new mother to leave such a little baby," she worries, "even though I know I'm leaving him in very competent hands."

- Meredith Wilson, at age 2, is racing into early childhood and the age when toddlers begin to establish independence and develop a sense of self. As Meredith struggles to draw a line—sometimes a battle line—between herself and others, mother Patty sighs, "It's hard to keep up with her." But keep up with her she does, for each time Meredith claims her independence she also wants to return to mom and re-establish the bond of basic trust that was the legacy of her infancy.

- Gilberto Agosto, at 3, is old enough to have his own cubbyhole at the day care center he attends in East Harlem, New York. Although he and his friends are only beginning to understand the differences between boys and girls, already they prefer to play with others of the same sex. One of the things Gilberto shares with boys his own age is a biological tendency to be more aggressive than girls.

- James McManus, at age 4, not only knows that he's a boy and acts like one, but he also is forming autobiographical memories that will both shape and reflect his self-identity. When he's an adult, perhaps his first memory of life will be of running faster than the wind as he plays a game of ghost with his mother.

- At an art class in Boston, children eagerly paint pictures depicting themselves, their families, and important events, such as the loss of a pet. According to art educator Nancy Smith, a child's drawings are "the roots of it all—the first emergings of a great enormous tree" that is the child's sense of self.

- It's the first day of school for 6-year-old Jamillah Johnson, who lives with her grandmother in the Boston suburb of Roxbury, Massachusetts. At the end of early childhood, Jamillah is about to take a big step into the world of teachers, classmates, and formal education. Walking to the school, Jamillah's grandmother offers her granddaughter some loving admonitions.

Now answer questions 1–4, pages 363–364.

PROGRAM TWO: CHILDHOOD AND ADOLESCENCE (Ages 6–20)

Program Two presents the stories of nine young people who spell out different versions of growth and development during a sometimes tumultuous season. The biological clock slows growth in childhood, giving humans a latent period in which to learn the skills and information critical to culture. According to the social clock, this is the time to go to school, to work at acquiring a sense of industry, and develop feelings of being useful and competent. At the close of childhood, the biological clock again ticks loudly, bringing on adolescence and what probably is the most challenging and complicated season of life. By the time it ends, adolescents have formed a fragile sense of who they are and have taken up authorship of their life stories.

STORIES

- Six-year-old Jamillah Johnson carries a pink backpack and a serious expression to her first day at kindergarten in her neighborhood elementary school. Her grandmother is holding her hand now, but school and schoolteachers are about to exert a very significant influence on her cognitive and social development. In class the teacher reminds Jamillah and her classmates: "Raise your hand. Stand straight. Stand still and quiet. Stand right behind the person in front of you." On the playground Jamillah makes overtures to a potential friend. Her grandmother's words to her ring true: "You're going to be on your own. Everything you learn now is important for <u>you</u>."

- Nine-year-old Karl Haglund learns to love his handicapped brother Gerry. Describing her sons' sometimes argumentative relationship, the boys' mother comments, "We had to explain to Karl that at times he has to be more understanding of Gerry's special needs. That can be frustrating for a 9-year-old to understand. Yet, when push comes to shove, Karl wants to be where Gerry is." Karl has a reading problem, but is nevertheless developing a sense of industry.

- Eleven-year-old Jason Kennedy takes his part on the family farm—raising calves, doing chores and working in the field. "Our kids are not smart 'streetwise' like kids in town," says his father Jim, "but you tell a farm kid to do anything and he'll give it a shot." Jason's love is showing animals. Though he has lost his share of competitions, he revels in first-place ribbons and his father's pride.

- Twelve-year-old Candy Reed finds her way amid the social stresses of junior high school. Her grandfather, who helps his single-parent daughter raise Candy, says fondly, "She wants to feel she's a big girl, but we don't feel she's quite big enough yet." Her mother worries that some day discipline may alienate her daughter. "You can't be too permissive," she says, "you can't always just be a teenager's friend."

- Fifteen-year-old Kim Henderson must be responsible for her 18-month-old daughter Angela at the same time she works to finish high school herself. Although her mother and sister care for the toddler during the day, Kim's after-school time is spent with her daughter. Kim is trying hard to be the best mom she can be—a blend of her own mother and grandmother—but she wants her daughter's experiences to be different from her own. "I hope Angela doesn't turn out like me," she says, "I don't want her to have to have a baby the same time I did. It could ruin her life."

- Seventeen-year-old Nuket Curran may be the most rebellious of the program's protagonists. Through her choices of music, clothing, and hair, she tries to distance herself from the crowd in her urban Pittsburgh high school. A talented artist, she has filled an autobiographical canvas with starkly drawn symbols of her emotions. Despite her rebelliousness, Nuket has some sound advice for adults, reminding them that, "Being a teenager is not easy. We're growing up and should be allowed to make stupid mistakes from time to time—to a point!"

- Eighteen-year-old Michael Shelton clashes with his stepfather over the family car. Unrepentant about his speeding, he jokes that his next investment will be a radar detector. Michael, who is about to enter a college program in commercial art, discusses his complex feelings about his biological parents and stepparents. Michael wants the girls he dates to be good looking. Although his dating habits have been tempered by the threat of AIDS, "Nobody in his right mind would go with a girl only for her personality and mind," he says.

- Eighteen-year-old Trey Edmundson gets ready for the senior prom. Although his mother died when he was 12, he has had a loving aunt and uncle looking out for him. "Thank God we went through that period quickly," says his uncle, Reverend Charles Stith, of Trey's adolescence. "At times," Trey admits, "teens want to be treated as adults; other times they want to do things kids would do."

- Seventeen-year-old Heather Robinson finds time to give love and attention to young children and an elderly woman. She volunteers to read stories to preschoolers at the local library. And for three years she has enjoyed a very precious relationship with Ida Rhine, an elderly, blind, retired musician she calls her advisor, confessor and confidante. Soon Heather will be going away to college, where she has been accepted into a prestigious writing program. What has helped her, she says, "is just knowing that I've actually done something for somebody and not always feeling like there's so much out there that's wrong and that there's nothing I can do."

Now answer questions 5–8, pages 365–366.

PROGRAM THREE: EARLY ADULTHOOD (Ages 20–40)

Program Three of *Seasons of Life* explores development during early adulthood. As children and teenagers, we impatiently await the "rites of passage" that officially signal our entry into this season. But for young adults today, the vast array of lifestyle choices may be intimidating as well as exciting. The social clock ticks very loudly for people in their 20's and makes enormous demands all at once. The messages are urgent: Get a job. Find a mate. Start a family. In early adulthood we form a dream of the future and try to make that dream a reality.

STORIES

- It's an important day in the life of Justin Miller, age 21. On the threshold of early adulthood, he prepares to graduate from Carnegie-Mellon University with a degree in artistic design and hopes for a promising career. "My biggest fear is that I won't be successful," he worries. "You can easily get stagnant in some place and not move anywhere. That would be the worst that could happen to me." Like many college graduates today, Justin is putting his personal life on hold until his career is successfully launched. Still, he admits to thinking a lot about having a family, especially at the end of the day when he returns to his empty apartment.

- May-Ling Agosto's life, at age 21, has never been easy. Raised by her mother in New York City's Spanish Harlem, May-Ling married at 15 and now has a 3-year-old son. She is separated from her husband. "We're very strong pigheaded women in our family," she smiles. "If a husband is not going to do anything for you, what do you need him for?" May-Ling has fought her way back from cocaine addiction and looks ahead with excitement. She is training as a computer technician and has just moved into a one-bedroom, subsidized apartment in the Bronx after a 3-year wait. "Now it's just up to me," she vows.

- Anthony and Julianne Cugini, 28 and 23, belonged to the same Catholic church all their lives before they met at a youth organization event. Religion and family are strong supports for them. The Cuginis have been married one year and have a baby girl. "My dreams aren't far-fetched at all," Julianne says. "I want to have a marriage that works." Her husband agrees, "I'm as happy as I've ever been in my life," he confides.

- Donna Radocaj, 31, feels her dreams have been shattered. Divorce has left her struggling to cope with her two small children, and in contrast to her comfortable married life, now there have been days with "no food and no money." Resolve and bitterness come together for Donna at this point in her life: "I thought by now I'd have my house in the country," she reflects. "I'm settling for second best."

- Phillip and Patti Wilson, both 30, were married 9 years before their daughter Meredith was born. They admit that marriage and parenting have revamped earlier dreams. "When I was 20," Patti says, "I never thought I'd be in Pennsylvania. I never thought I'd be married. I *never* thought I'd have kids." Phillip confesses he had a TV image of what a husband was. "When the husbands came home the wives were relaxed." Despite the fact that their marriage has had its ups and downs, Patty speaks poetically about their baby: "I feel like I found something that I had before, but I just didn't know it. Like a new room in the house. You've passed the door so many times, but you never bothered to open it. Then I opened it, and it's the best room in the house."

- Thirty-one-year-old Bronwyn Reed was an unwed mother at 19. Now she supports her daughter Candy with a job she enjoys, but is not committed to. She is a physical therapist at a local hospital. "I don't have a career I can be proud of," she says in disappointment, "but when you have a job you don't want to do, you have to seek fulfillment in other ways." For Bronwyn that is music; she is an accomplished vocalist. Bronwyn feels ambiguous about her status as a single mother. "Sometimes I'm glad I'm not married," she says, "and sometimes I'd give an arm and a leg to have somebody else to talk to."

- Free-lance writer David Nimmons and union organizer David Fleischer are an openly gay couple in their early thirties. Like many adults his age, Nimmons was ready to settle down when he and Fleischer met. "People I know who are in couples are really working at something that seems like a good and valiant challenge," he says. "How lucky I am to have found this, because it fits so well for me."

- The idea that she may never marry and have children is a painful one for Christine Osborne, 39. While she is a very successful Chicago ad executive, personally she longs for a family. "I feel very envious of people who not only have happy marriages, but who have children," she says. Ruefully she notes her grandmother's dying words to her: "You're too picky." "It hasn't happened," she shrugs, tears in her eyes. "Now what?"

- Deborah and Charles Stith, both in their late thirties, would seem to have it all: He is an ordained Methodist minister in Boston. She is a physician and the State Commissioner of Public Health. They have two small children Charles calls "major events," and Deborah jokes about wanting seven more. Yet they too admit to struggling. They also raised Deborah's now-teen-age nephew after her sister died. "There are many points along the way we could have gotten divorced," Deborah acknowledges. "I'm not going anywhere," says Charles. "I picked the best I'm capable of picking."

Now answer questions 9–12, pages 367–368.

PROGRAM FOUR: MIDDLE ADULTHOOD (Ages 40–60)

Program Four explores development during the fifth and sixth decades of life. Gray hair, wrinkles, and reading glasses are signals from the biological clock that life is half over—that there is more "time lived" than "time left." This is the season in which the biological and social clocks fade in importance and the psychological clock ticks more loudly. For those who have reached positions of expertise and power, middle age will be a time of personal and social command. For others who have experienced great losses—that of a child, of a spouse, or of a job—it will demand new directions. It is a season that ushers in androgyny: Women find new strength and men become more nurturing. Caring for others— what Erikson called generativity—is also emblematic of the middle years, as both sexes provide help to their extended families and the larger "human family," and begin to create a legacy for those who will come after them.

STORIES

- Former Clevelend Browns quarterback Brian Sipe faces the "precipice" of middle age at 37—after a lifetime of sports. "We all wish we were kids a little longer than we were," he says, "but I'll always be proud of what I did." Sipe is anxious to re-direct his future. He thinks he might like to be an architect. And although he misses the regimen of professional sports, he's ready to take a next step. "There's nothing stopping me," he declares.

- Born January 1, 1946, Kathleen Wilkins was America's first official "baby boomer" and is now the first of her cohort to reach middle age. Recently separated from her husband, Kathleen has returned to college to study for an MBA while continuing her career in restaurant management. Not afraid of growing older, she says, "At 40, as a 'boomer,' I have more to offer than a 22-year-old. I can bring not only an expertise that I went back and studied for, I can bring life experiences."

- Daniel Cheever, 44, is the president of Wheelock College in Boston. At one time he considered leaving the field of education, but after 10 years as a public school superintendent, "I decided I had paid my dues in one field and that's where I would reap my reward." Cheever and his wife Abby both have professional careers, and Cheever has done most family chores since Abby entered law school more than 10 years ago. Both he and Abby look forward to the challenge of their life together now that their children will be in college.

- Jim Kennedy, 45, is celebrating his 25th wedding anniversary on the family dairy farm Jim's father mortgaged to him two decades ago. Although he and wife Rita once considered going out of business and can remember struggling to find 24 cents to buy a loaf of bread, today they sell 1.25 million pounds of milk a year. Jim has accumulated enough acreage to be able to offer any or all of his five children part of the family business. "The money isn't in farming," he says, "but it's a darned good life."

- The good life is gone for Matt Nort, 52, who lost his job at a Pittsburgh steel mill some time ago. Nort, who does not have a high school education, gets by on part-time work and hopes for a chance. "The dream is gone, probably gone forever," he worries. As he struggles to reconnect his life, he confronts many new feelings, including depression, bewilderment, and even thoughts of suicide. Yet Nort has strong faith, and a loving family. "It's just you and me against the world," his wife says, smiling at him.

- Kris Rosenberg, 55, fought her way back from a crippling bout with polio, a divorce, and perhaps most important, a devastating loss of self-esteem. Today she directs "Returning Women," a training program for older women who want to re-enter the work force. "I can be real to them," she explains, noting that she was once as frightened and tentative as her students. Kris and her husband have reunited since her comeback, but they maintain separate lives, checking accounts, and rooms in the house. Having in effect, turned her losses into gains, Kris notes, "For the first time I feel like an individual person."

- Harriet Lyons, 59, climbed out of despair and totally reconstructed her life after losing everything, including family mementos, when she lost her job. Eventually she found work with the Bank of Boston, and is setting goals for herself once again. Lyons raises her granddaughter Jamillah because her daughter, Jamillah's mother, is addicted to drugs. Despite this sadness, she is excited about the friendships she is strengthening with her six children and grandchildren, noting, "I guess in a lot of ways I'm coming to where I wanted to be in the first place. I feel good about myself. I've got a lot to offer."

- Dave and Ruth Rylander, 57 and 53, have lived a parent's nightmare. Their daughter Lynn died of leukemia when she was 19. As a result of their love for her and their own longing for meaning in their lives, both Rylanders have taken up careers in social service. Ruth runs peace movement activities for the Presbyterian Church and Dave is the director of a local food bank. "I think she'd like what I'm doing today," Dave says. "She was the kind of young woman who cared about people." Ruth speaks eloquently about their lives today: "We've seen tragedies, joys. We have done our apprenticeship. Sometimes I am more cautious, but in many ways I'm less cautious. After all, what do I have to be afraid of?"

Now answer questions 13–16, pages 369–370.

PROGRAM FIVE: LATE ADULTHOOD (Ages 60+)

Late adulthood is a season of great diversity. It is the season upon which the 20th century, with World Wars I and II, the Great Depression, and a 25-year increase in life expectancy, has had its most telling impact. Late adulthood is one kind of season if you have your health and financial resources; it is quite another if you don't. Some older adults extend the activities of middle age into their 70s. Others take advantage of their newly found leisure to change their lives completely. As the biological clock eventually runs down and the end of life comes into view, individuals work to achieve a sense of integrity. Now the lifelong suspense leaves their story, and they see how it will end. And they wonder: Was it a good story? Was it true to who I was? And are there a few last things I might do to give my life a better ending?

STORIES

- Harry Crimi, 61, has operated the family butcher shop in south Philadelphia for as long as he has been married—40 years. His wife Antoinette, 60, was born in the house they live in now, and her 93-year-old father is still with her. "He deserves the honor and glory of dying at home," she says. Harry and Antoinette dote on their family. Their children and grandchildren visit often. "You don't have time to cater to your children," Harry explains, "but you make time for your grandchildren." Where did the time go? "When you're 20 or 30," Harry shakes his head, "you don't imagine being 61 years old. But now that I'm 61, I wonder where all the years went."

- In Detroit, Tom and Vivian Russell, 61 and 60, are buying a new home in retirement. He has worked 40 years for the post office; she has been a social worker. Tom grew up in a black family in the segregated South where he remembers life as a constant struggle. Vivian was also raised in a poor family, one of 9 children. They are a remarkably giving couple. Tom meets in Christian fellowship with inmates of a Federal penitentiary and Vivian is training as a hospice volunteer. As Tom says, "I hope we can continue to help people, continue to show our warmness, continue to do it together."

- Seventy-four-year-old Lyman Spitzer will leave an important scientific legacy. NASA will soon put into space a telescope designed by the astronomer/physicist. Spitzer remains fascinated with knowledge and continues the vigorous physical activities he has enjoyed all his life, not the least of which is mountain climbing. "No one has reproached me for being overly rash at my age," he contends. Spitzer has been married 51 years to Doreen Spitzer, who heads the American School in Greece. After all he and his wife have studied and experienced, life is still a fascinating puzzle. "I can well imagine," Spitzer says, "long after we know all that we need to know about the universe, we may still be trying to understand the nature of life."

- Milton Band, 78, and his wife Rowena, enjoy the "good life" at a retirement center in Holiday Springs, Florida, where the relaxed atmosphere has afforded them a new intimacy in their marriage. "I feel like I'm 16 years old," Rowena says. "It's a wonderful, unbelievable thing (retirement), all of it, like a dream." Milton, who left his law career nine years ago, celebrates time to himself and an escape from the rat race.

- Francis Kennedy, 70, jokes about a few aches and pains and a forgetful moment or two, but he is still active on the family farm, baling hay and milking cows. "You could sit down, and I could sit down, and we could really cry the blues with each other. That wouldn't do any good," he claims. "We've got to get going. Tomorrow's another day. Be ready for it."

- Ellen Hanes is not as fortunate as her peers in good health. She is one of the frail elderly who requires constant care, and she lives in a private nursing home in Flushing, New York. Thirty-five percent of all sick elderly Americans will spend their life savings in a nursing home before they die. "If I had known I would be like this, I would have made more arrangements for myself. The doctors put me here," she says.

- Miriam Cheifetz lost her husband a year ago. When she found her loneliness unbearable, she moved to a group home in Chicago where she feels more comfortable. The women in the home share facilities, chores, and the stories of their lives. Miriam treasures memories of her husband and frequent visits from her affectionate family: "There's a thread that goes through your whole life, and when you see some of that in your children, you feel that your life was not in vain."

- "I hope I don't lose my upper plate when I blow," is Minna Citron's comment as she blows out the candles on her 90th birthday cake. After a long, unconventional life (Her ex-husband once said, "Why can't you be like other people and just stay put?"), Citron enjoys her position as "curator of a career"—her own, as a prominent artist. Mobbed by friends at her birthday party, she is in high spirits: "I love my friends . . . I want to go on enjoying life if I can."

- George Nakashima, 84, chose a life of contemplation and art. Although he began work as a railroad gandy dancer in the Northwest, a career in architecture and design eventually took him to Paris, India, and Japan. During World War II, Nakashima, his wife, and 6-week-old daughter were forced into a detention camp set up by the U.S. government. There he began to train under a fine Japanese carpenter, schooled in the traditional manner. Nakashima came to love working with wood and developed a spiritual affinity for nature. Today he and his son and daughter painstakingly fashion works of art. "I take great pride in being able to take living things that will die, and give them a second life. It is a great feeling to be a part of nature, and to be a part of life itself," he says.

Now answer questions 17–25, pages 371–377.